Abiotic Disorders of Landscape Plants

A Diagnostic Guide

Laurence R. Costello
University of California
Cooperative Extension Environmental Horticulture Advisor
San Francisco–San Mateo Counties

Edward J. Perry
University of California
Cooperative Extension Environmental Horticulture Advisor
Stanislaus County

Nelda P. Matheny
HortScience, Inc., Pleasanton, CA

J. Michael Henry
University of California
Cooperative Extension Environmental Horticulture Advisor
Riverside and Orange Counties

Pamela M. Geisel
University of California
Cooperative Extension Environmental Horticulture Advisor
Fresno County

University of California
Agriculture and Natural Resources

Publication 3420

Kitty Walker

To order or obtain ANR publications and other products, visit the ANR Communication Services online catalog at http://anrcatalog.ucanr.edu/ or phone 1-800-994-8849. You can also place orders by mail or FAX, or request a printed catalog of our products from

University of California
Agriculture and Natural Resources
Communication Services
1301 S. 46th Street
Building 478 - MC 3580
Richmond, CA 94804-4600
Telephone 1-800-994-8849
510-665-2195
FAX 510-665-3427
E-mail: anrcatalog@ucanr.edu

Publication 3420
ISBN 1-879906-58-9
Library of Congress Control Number: 2002107829
Third printing, 2014

Photo credits appear on pages v–vi, which constitute an extension of this copyright page.

UC PEER REVIEWED

This publication has been anonymously peer reviewed for technical accuracy by University of California scientists and other qualified professionals. This review process was managed by ANR Associate Editor for Pest Management.

Printed in Canada on recycled, acid-free, paper

3m-rep-2/14-SB/CR

Contents

Kathy Walker

Acknowledgments

The authors gratefully acknowledge the Western Chapter of the International Society of Arboriculture and the Pesticide Applicators Professional Association for providing funds to support the development of this publication.

Contributors, Technical Advisors, and Reviewers

The authors sincerely thank the following individuals for their valuable contributions to this publication. Some supplied technical content while others served as reviewers or advisors. We are particularly indebted to those who made substantial contributions of information (in bold). Collectively, these individuals have helped to ensure the accuracy and completeness of this publication.

Colin Bashford, CBA Ltd., Twyford, Hampshire, UK

Alison M. Berry, University of California, Davis

Kenneth Blonski, University of California Forest Products Laboratory, Richmond

Jennifer Bowman, HortScience, Inc., Pleasanton, CA

Edward Brennan, HortScience, Inc., Pleasanton, CA

David W. Burger, University of California, Davis

Gary Chan, University of California Agriculture and Natural Resources Analytical Laboratory, Davis

James R. Clark, HortScience, Inc., Pleasanton, CA

Scott Cullen, Greenwich, CT

James A. Downer, University of California Cooperative Extension, Ventura

Steve H. Dreistadt, University of California Statewide Integrated Pest Management Program, Davis

Clyde L. Elmore, University of California Cooperative Extension, Davis

Richard Y. Evans, University of California Cooperative Extension, Davis

Mary Louise Flint, University of California Statewide Integrated Pest Management Program, Davis

David A. Grantz, University of California Kearney Agricultural Center, Parlier

Bruce W. Hagen, California Department of Forestry, Santa Rosa

Paul Hanson, Perthshire, Scotland

Richard W. Harris, University of California, Davis

Keiron Hart, Woking, UK

Gary W. Hickman, University of California Cooperative Extension, Mariposa

Donald R. Hodel, University of California Cooperative Extension, Los Angeles

Katherine S. Jones, University of California Cooperative Extension, Half Moon Bay

John M. Lichter, Tree Associates, Winters, CA

Clifford Low, Perry Laboratory, Inc., Watsonville, CA

James D. MacDonald, University of California, Davis

Robert W. Miller, University of Wisconsin, Stevens Point

Stuart G. Pettygrove, University of California Cooperative Extension, Davis

Dennis R. Pittenger, University of California Cooperative Extension, Riverside

Stephen L. Quarles, University of California Forest Products Laboratory, Richmond

Thomas B. Randrup, Danish Center for Forest, Landscape and Planning, Hoersholm, Denmark

Michael Santos, HortScience, Inc., Pleasanton, CA

Lawrence J. Schwankl, University of California Cooperative Extension, Davis

Roger H. Shaw, University of California, Davis

John Shelly, University of California Forest Products Laboratory, Richmond

E. Thomas Smiley, F. A. Bartlett Tree Research Laboratory, Charlotte, NC

Richard L. Snyder, University of California Cooperative Extension, Davis

Lucy Tolmach, Filoli Center, Woodside, CA

Jane Whitcomb, HortScience, Inc. Pleasanton, CA

Production

Technical Review Coordinator

Mary Louise Flint, University of California Statewide Integrated Pest Management Program, Davis

Image Production

Jack Kelly Clark, University of California Agriculture and Natural Resources Communication Services, Davis

Steve Lock, University of California Agriculture and Natural Resources Communication Services, Davis

Project Editor

Stephen W. Barnett, University of California Agriculture and Natural Resources Communication Services, Davis

Design and Layout

Celeste Rusconi, University of California Agriculture and Natural Resources Communication Services, Davis

Indexing

Ellen Davenport, Sebastopol, CA

Photo Credits

Bonnie Lee Appleton, Virginia Cooperative Extension, Virginia Beach, VA: figs. 5.123, 5.125. Wesley Asai, Agricultural Consultant, Turlock, CA: figs. 5.43, 5.153, 5.162, 5.168. California Air Resources Board, Sacramento: figs. 5.114, 5.115. Jack Kelly Clark, UC ANR Communication Services: figs. 1.8, 2.1, 4.11a, 4.17 (a,b), 5.21 (a,b). Laurence R. Costello, UC Cooperative Extension: figs I.1, I.3, I.4, 1.1, 1.2, 1.3 (a,b,c), 1.4, 1.5, 1.7, 1.9, 1.10, 1.11, 1.12a, 1.13 (a,b), 1.14, 3.1, 3.7, 3.13, 3.15, 3.19, 3.25, 4.1b, 4.2 (a,b), 4.3, 4.5, 4.6 (a,b), 4.10, 4.11b, 4.20 (a,b), 4.21, 5.1 (a,b), 5.2, 5.5, 5.6, 5.7, 5.8, 5.9, 5.10, 5.11, 5.12, 5.13, 5.14, 5.15, 5.16, 5.17, 5.18, 5.19, 5.20, 5.21c, 5.22, 5.23, 5.24 (a,b,c), 5.26 (a,b), 5.27, 5.28 (a,b), 5.29 (a,b), 5.31, 5.32 (a,b), 5.33, 5.34, 5.35 (a,b), 5.46, 5.54, 5.63 (a,b), 5.64 (a,b), 5.65, 5.66, 5.67, 5.68 (a,b), 5.70, 5.71, 5.72, 5.73 (a,b), 5.74, 5.75, 5.76 (a,b), 5.79 (a,b), 5.81, 5.82, 5.85, 5.86, 5.91 (a,b), 5.92, 5.93, 5.95, 5.97 (a,b), 5.98, 5.99, 5.100 (a,b), 5.102, 5.103 (a,b), 5.104 (a,b), 5.105 (a,b), 5.106, 5.107b, 5.108 (a,b), 5.109 (a,b), 5.110, 5.111, 5.124, 5.126

(a,b), 5.127, 5.128, 5.131, 5.134, 5.135 (a,b,) 5.136 (a,b), 5.140 (a,b), 5.141, 5.142, 5.143, 5.144 (a,b), 5.146, 5.147, 5.170 (a,b,c), 6.7, 6.8, 6.9, 6.10, 6.11, 6.12, 6.14, 6.15, 6.16, 6.17, 6.18, 6.19, 6.21. Clyde L. Elmore, UC Cooperative Extension: figs. 5.148, 5.149, 5.161, 5.163. Susan Frantz, Apogee Instruments Inc., Logan, UT: fig. 5.101. Pamela M. Geisel, UC Cooperative Extension: figs. 5.41, 5.77, 5.80, 5.83, 5.87, 5.133, 5.137, 5.138. David Grantz: figs. 5.112, 5.113, 5.117, 5.118, 5.119. J. Michael Henry, UC Cooperative Extension: figs. 5.96, 5.107a, 5.129 (a,b), 5.169. Donald R. Hodel, UC Cooperative Extension: figs. 5.37, 5.38, 5.39, 5.44, 5.45. HortScience, Inc., Pleasanton, CA: figs. 1.15, 3.2, 3.3, 3.4, 3.5, 3.6, 3.10, 3.11, 3.14, 3.16, 3.18, 3.20, 3.22, 3.23, 3.24, 3.26, 4.4, 4.7, 4.8b, 4.13, 5.25, 5.42, 5.47, 5.48, 5.50 (a,b), 5.51, 5.52, 5.53, 5.55, 5.57, 5.58, 5.59 (a,b), 5.60, 5.61, 5.89, 5.90, 5.132, 5.156. John M. Lichter, Tree Associates, Winters, CA: fig. 6.20 (a,b). Joseph McNiel, Consulting Arborist, Pleasant Hill, CA: fig. 5.88. Integrated Pest Management Program, Univ. of California, Davis: figs. I.2, 1.6, 3.8, 3.9, 3.12, 3.17, 3.21, 3.27, 4.18, 5.30, 5.36, 5.116, 5.164. Suzanne Paisley, formerly of UC ANR Communication Services: figs. 2.2, 2.3, 2.4, 2.5, 2.6, 2.7, 2.8, 2.9, 5.40, 6.13. Edward J. Perry, UC Cooperative Extension: figs. 1.12b, 4.1 (a,c), 4.8a, 4.9 (a,b), 4.12 (a,b), 4.14, 4.15, 4.16, 4.19, 5.3, 5.4, 5.56, 5.62, 5.69 (a,b), 5.78, 5.84, 5.94, 5.145, 5.150, 5.151, 5.152, 5.154, 5.155, 5.157, 5.158, 5.159, 5.160, 5.165, 5.166, 5.167, 6.1, 6.2, 6.3, 6.4, 6.5, 6.6. Thomas B. Randrup, Danish Center for Forest, Landscape and Planning, Hoersholm, Denmark: fig. 5.49b. E. Thomas Smiley, F. A. Bartlett Tree Research Laboratory, NC: figs. 5.120, 5.121 (a,b), 5.122.

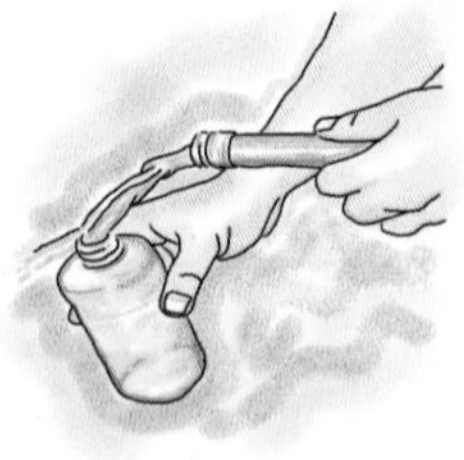

Abiotic Disorders of Landscape Plants

A Diagnostic Guide

Introduction

Landscape plants can be injured by biotic and abiotic agents. Abiotic or nonliving agents include environmental, physiological, and other nonbiological factors (fig. I.1). Biotic agents are living organisms such as insects, pathogens, nematodes, parasitic plants, and viruses (fig. I.2). This publication focuses on abiotic disorders such as water deficits, aeration deficits, nutritional deficiencies, specific ion toxicities, pH-related problems, and herbicide injury.

Whether the cause is biotic, abiotic, or both, an accurate diagnosis is almost always needed to remedy the ailment (fig. I.3). This publication describes the strategies, techniques, and tools used in diagnosing plant problems, and it provides illustrations and guidelines to aid in diagnosing abiotic disorders. An excellent source of information on biotic agents and disorders is *Pests of Landscape Trees and Shrubs: An Integrated Pest Management Guide* (Dreistadt 1994).

Much of this publication is devoted to the symptoms, occurrence, and diagnosis of specific abiotic disorders. Injury symptoms that mimic a particular abiotic disorder are identified ("look-alike disorders"), and lists of sensitive and tolerant species are in-

Figure I.1. Abiotic disorders are caused by nonliving agents. Here, marginal necrosis and pitting on a leaf of prospector elm (*Ulmus wilsoniana* 'Prospector') are symptomatic of boron toxicity, an abiotic disorder.

Figure I.2. Biotic disorders are caused by living agents such as insects and pathogens. Here, redgum *(Eucalyptus camaldulensis)* leaves covered with "lerps" of the redgum lerp psyllid *(Glycaspis brimblecombei)* indicate a biotic disorder.

cluded. Color photographs and easy-to-use summary tables describe symptoms and help identify causes. Although some information on remedies is provided, treatment is not the focus of this guide.

The guide also contains:

- guidelines for collecting soil, water, and tissue samples and interpreting test results
- descriptions of naturally occurring plant characteristics that are commonly mistaken as disorders
- case studies of actual problems, including their diagnoses and treatments
- a glossary and an index

Collectively, this information will assist landscape professionals in accurately diagnosing abiotic disorders and thereby maintaining landscape health (fig. I.4).

Figure I.3. An accurate diagnosis is needed to remedy most disorders. This arborvitae (*Thuja* sp.) is showing uniform discoloration and decline of the canopy. These symptoms may have been caused by biotic agents (e.g., a root pathogen) or by abiotic agents (e.g., water deficit, aeration deficit, mechanical injury, or gas line leak).

Diagnosis is the process of identifying the cause (or causes) of a problem. To determine the cause of injury, plant diagnosticians collect and evaluate information on symptoms, site conditions, and plant history.

Figure I.4. Healthy and attractive landscapes can be maintained if disorders are accurately diagnosed and appropriately treated.

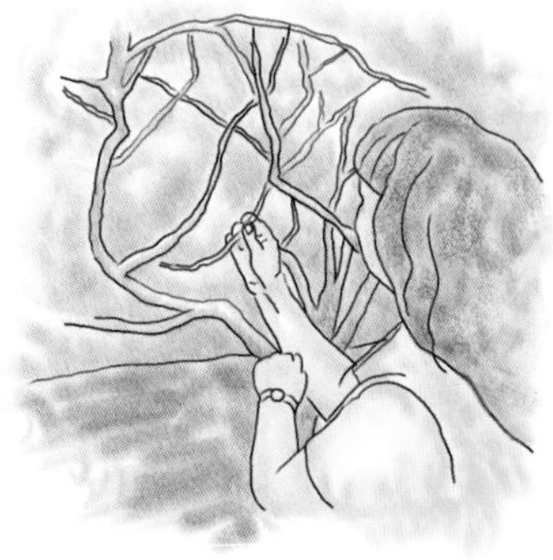

CHAPTER 1

Diagnosing Abiotic Disorders

Challenges, Traits, Strategies, and Tools

Accurate and timely diagnosis of plant problems is an essential practice in sustaining healthy and resource-conserving landscapes. Biotic and abiotic disorders are common, even in the best-managed landscapes. Failing to accurately identify and quickly control problems often decreases landscape quality and increases maintenance costs.

Not only is problem diagnosis very important, it is also very challenging: it may be one of the most difficult practices in landscape management. Successful diagnosis requires an integration of horticultural knowledge, experience, and problem-solving skill (fig. 1.1). It may take many years before a high level of competence is achieved. To begin, however, it is important to understand the challenge of diagnosing landscape problems and then learn about diagnostic traits, strategies, and tools.

Figure 1.1. Diagnosis of plant problems can be very challenging. Hundreds of species are planted in landscapes, site conditions vary tremendously, the plant environment changes, multiple factors may be involved, and injury symptoms can be subtle.

The Diagnostic Challenge

The diagnosis of landscape problems is challenging for several key reasons:

- Many biotic and abiotic agents can cause injury.
- Landscapes exhibit tremendous variability.
- A problem can have multiple factors or causes.
- Chronic problems may express subtle symptoms.

Biotic or Abiotic Causes

Biotic and abiotic disorders are numerous and diverse. Many insects, pathogens, nematodes, environmental factors, and physiological factors can cause problems for plants in the landscape. Knowing many of the possible causes and learning their identifying characteristics can be a challenge, and simply distinguishing between biotic and abiotic causes can be difficult (see "Distinguishing between Biotic and Abiotic Disorders," p. 5).

Figure 1.2. Landscapes vary considerably in species composition. In this commercial property, an assortment of groundcovers, shrubs, and trees are planted. The number of species and the variation in species found in landscapes can make diagnosis difficult.

Landscape Variability

Landscapes not only vary widely in species, soils, and environmental conditions, they also change over time.

- Hundreds of species are planted in landscapes, and each landscape has a different combination of species. Effective landscape management and successful problem diagnosis require identification of each species and some knowledge of its culture (fig. 1.2).
- Planting sites and conditions vary tremendously: soils, terrain, microclimate, and exposure can be quite different within regions and even within local areas. No two planting sites are exactly alike. In fact, conditions may be very different for plants just a few feet from each other (fig. 1.3).

Landscapes are dynamic systems—the plant environment changes from day to day. These changes may be favorable or unfavorable; and plants change anatomically, morphologically, and physiologically as they grow and mature. Changes in the environment and in plant development make diagnosis even more challenging.

Other Considerations

- In many cases, multiple factors may be involved in a disorder (fig. 1.4). For example, soil moisture deficit, air temperature, wind, disease, and chemical treatments may injure plants by themselves or they may collectively contribute to symptoms of injury.
- Chronic injuries are common, can be subtle, and are often difficult to diagnose. For example, low to moderate levels of salt in the soil can cause a species to grow slowly, and other symptoms associated with excess soil salts, such as leaf chlorosis and marginal necrosis, may not be expressed.
- Generally, few tools are available for evaluating plant disorders (see "Tools for Diagnosis," p. 10). The available tools are very basic and of limited utility. Although laboratories can analyze samples, little research is available to help interpret the data. Often, diagnosticians must rely on their own observations, education, and experience to evaluate injuries and diagnose disorders.

Distinguishing between Biotic and Abiotic Disorders

A very important step in diagnosing plant problems is to determine whether the cause is abiotic (nonliving) or biotic (living).

Biotic factors include pathogens (fungi, bacteria, viruses, mycoplasmalike organisms, mollicutes), nematodes, insects, mites, mollusks, vertebrates (e.g., rodents, birds), and parasitic plants (e.g., mistletoe). Abiotic factors include physical or environmental problems (e.g., moisture extremes, temperature extremes), mechanical problems (e.g., root cutting), and chemical problems (e.g., high soil salinity, phytotoxic pesticides).

Determining whether an injury is caused by an abiotic or biotic factor can be difficult, because symptoms associated with biotic and abiotic disorders can be similar. For example, water deficit and vascular disease can cause similar symptoms: marginal

Figure 1.3. Soil conditions vary in landscapes. At a college campus, shown here, a trench being dug for a utility line uncovers different soils over a short distance (A). The soil in the foreground has a sandy texture (B), while the soil in the background has a considerable amount of construction debris (C). Plant performance in the two soils would not be expected to be the same.

necrosis, chlorosis, leaf drop, and dieback (fig. 1.5). In many cases, an examination of soil properties, environmental conditions, and plant tissues may be needed to determine the cause.

The following characteristics may help differentiate between biotic and abiotic disorders. Keep in mind that these are **general** characteristics and exceptions do exist.

Characteristics of Biotic Disorders

- Physical evidence may be present when a disorder is caused by a biological organism. For example, insect pests may be visible on the plant, or cast skins or frass may be found. With fungal diseases, white patches on leaves (powdery mildew), spore masses (rusts), or mushrooms (oak root fungus) may be seen (fig. 1.6).
- Biotic injury may spread progressively throughout a plant and onto other plants of the same species. For example, fungal leaf spots may spread quickly through a plant's canopy during cool, rainy weather in spring.
- Many diseases and insect infestations are limited to a particular species. For example, foliar injury caused by powdery mildew on London plane *(Platanus × acerifolia)* will not be found on other species in the area. Keep in mind, however, that some biotic agents have wide host ranges and may affect more than one species (e.g., fireblight affects several members of the rose family).

Characteristics of Abiotic Disorders

- Physical evidence of an abiotic disorder is usually not found on the plant. For example, boron toxicity, iron deficiency, aeration deficits, and wind produce injury symptoms but do not leave signs of the cause (fig. 1.7).
- Abiotic damage may or may not develop progressively throughout a plant (depending on the cause), and it usually does not move from plant to plant. For example, drift of a contact herbicide affects only leaves present at the time of drift, and new leaves will be normal.

- Abiotic damage may affect numerous plant species in a planting. For example, high soil salt levels or water deficits may affect all species present.

Primary and Secondary Causes

Abiotic factors often weaken plants, making them more susceptible to biotic factors. For example, overwatering may reduce the oxygen available to roots and predispose plants to root disease. Also, multiple abiotic factors may cause plant decline, leading to secondary biotic invasions. For example, wood boring insects are often attracted to sunburn-injured bark on drought-stressed trees. Evaluating all injury symptoms and determining whether the primary cause is biotic or abiotic are important steps in the diagnosis and management of plant problems.

Traits of a Good Diagnostician

A good diagnostician must combine many traits, including a strong educational background, field experience, and problem-solving abilities.

Knowledge of Plant Biology and Horticulture

A diagnostician must be able to identify local plant species and varieties and must understand their biology, growth habits, and normal appearance. This knowledge will come from several disciplines, including botany, plant physiology, soil science, hydrology, horticulture, entomology, plant pathology, ecology, climatology, and weed science.

Knowing the plant and its normal growth characteristics is critical to accurate diagnosis. Some plants have natural characteristics that may resemble abiotic disorders (see chapter 4). Or, a species may develop disorders simply because it is not adapted to local environmental conditions (for example, most juniper species are sensitive to flooded or overly wet soil conditions and may develop root disease).

Understanding of Landscape Cultural Practices

Because cultural practices play a critical role in the health of landscape plants, diagnosti-

Figure 1.4. Multiple factors are often involved in plant disorders. These London plane *(Platanus × acerifolia)* trees exhibit symptoms of foliar necrosis, branch dieback, and very slow growth. A combination of water deficit, wind, salinity, and high-density soils are producing these symptoms.

cians must have a thorough understanding of soil and water management, pest management, and other cultural practices. Poor irrigation practices—too much or too little water—account for many abiotic disorders. For example, dry soil conditions are often related to reduced plant growth and vigor, as well as to secondary injury from insect borers. Equally important is knowledge of the physical and chemical properties of soil and how those properties affect plant health. For example, plants growing in saline soil may develop symptoms similar to those caused by drought or pesticide phytotoxicity.

Access to Diagnostic Resources

It is very difficult to rely solely on memory for facts relating to plant diseases and disorders. Diagnosticians must be aware of and have access to resources to aid in diagnosis. Publications on entomology, plant pathology, and horticulture can be valuable tools (fig. 1.8) (see the references at the end of this book). Additionally, take advantage of resources such as UC Cooperative Extension publications and county soil survey maps, available at your local UC Cooperative Extension county office. The Internet makes many important resources available and also allows access to other experts and practitioners in the field, making the computer an increasingly important tool. Useful sites include the UC Integrated Pest Management Web site (http://www.ipm.ucdavis.edu/) and the California Irrigation Management Information System Web site (http://www.cimis.water.ca.gov/). Maintaining contact with other professionals in the field allows you to take advantage of their valuable experience.

Figure 1.5. Symptoms of certain biotic and abiotic disorders are similar. (A) For example, dieback in this Japanese maple *(Acer palmatum)* is caused by Verticillium wilt *(Verticillium dahliae)*, but symptoms are similar to water deficit injury. (B) Dieback in this sawleaf zelkova *(Zelkova serrata)* was caused by a water deficit, but symptoms are similar to Verticillium wilt injury.

Figure 1.6. Biotic disorders may show signs of the causal agent. On this rose (*Rosa* sp.) leaf, rust *(Phragmidium disciflorum)* is clearly visible.

Figure 1.7. Typically, abiotic disorders do not provide physical evidence of the causal agent. For example, the only evidence of high light injury in this dwarf periwinkle *(Vinca minor)* planting is a difference in leaf color between leaves exposed to full sun (chlorotic) and those in the shade (not chlorotic).

Working Knowledge of Current Diagnostic Techniques

Diagnosticians must know when and how to collect soil, plant, and water samples for laboratory analysis. For example, a soil analysis may be necessary to confirm suspected salt injury; a leaf analysis may be necessary to confirm a nutrient deficiency or to detect harmful pesticide residues. The reliability of laboratory analyses depends on the sampling procedure used. For a good analysis, samples must be taken and handled properly (see chapter 2).

An Inquiring Mind

Diagnosing plant problems often involves extensive detective work, especially when multiple factors are involved or when the cause of injury occurred weeks or months earlier. For example, root cutting may kill branches many months after it occurs; it may be difficult to diagnose the cause from the available signs. The ability to ask appropriate questions of people who may know the history of the site is critical (fig. 1.9). However, a diagnosis may need to be based on incomplete information or on information given by a person unskilled in horticulture. For example, a person unaware of the function of roots may not realize the impact of root cutting on tree health and may not volunteer the information during an interview.

Keen observation should lead the diagnostician to ask about activities that may have resulted in root injury. Or, the person being interviewed may be reluctant to reveal the necessary information, especially if the person accidentally caused the plant injury. For example, in cases of pesticide phytotoxicity, it may be difficult to find out which and how much pesticide has been used.

An Open Mind

An effective diagnostician keeps an open mind when faced with a plant problem, gathering as much information as possible and considering all possible options before making a decision (fig. 1.10). This reduces the chance of making an incorrect diagnosis based on a first impression of the problem. In many cases more than one problem is responsible for poor plant performance, a primary problem and a secondary one. A secondary problem may be the most obvious and relatively easy to diagnose. For example, the presence of borers in a declining tree may be apparent, but the primary problem, such as chronic drought stress, might be more difficult to detect. In this case, diagnosing borers as the cause of the tree's decline would lead to an incomplete and likely ineffective treatment of the problem. Also, keeping an open mind allows you to consider new pests and diseases, which, while rare, may still occur.

Ability to Think Logically

Along with keeping an open mind, it is important to think logically when making a diagnosis. Work from the simple to the complex by first considering common problems, such as too little or too much water, to the uncommon, such as unusual diseases.

Patience

Take enough time to do a thorough investigation, especially when multiple problems are involved. Don't rush into making a snap decision. Be prepared to defer a diagnosis to a time when you have collected and fully considered all relevant information. As with other professions, the more experience you have, the more successful you are likely to be. Developing and sharpening diagnostic skills requires study, staying abreast of new techniques and technologies, and a great deal of practice. Armed with sufficient knowledge, experience, and a strong desire to solve problems, you can become a successful plant problem diagnostician.

Figure 1.8. Diagnosis requires access to information resources, particularly reference manuals.

Tools for Diagnosis

Keep the following tools handy when investigating plant problems. Although you may not need all of these tools for every diagnosis, having ready access to them will save time and allow you to concentrate on the problem at hand.

Soil and Water Inspections

- shovel to examine soil for moisture content, collect soil samples, and to check roots for disease symptoms
- soil tube (12 to 24 inches, or 30 to 60 cm) to examine soil for moisture content, collect soil samples, and check soil for compaction
- portable pH and EC meters for quick pH and total salt analysis
- screwdriver to check soil for surface compaction
- sharpened steel rod (probe), ⅜ inch (9.5 mm) in diameter and 12 inches (30.5 cm) long, to check soil for compaction
- paper bags for soil samples
- small jars or plastic bottles with tight lids for water samples
- ice chest with "blue ice" for nematode samples

Plant Inspections

- pocket knife
- ice pick for probing bark
- steel rod, ⅛ inch (3 mm) in diameter by 3 feet (90 cm) long, to probe large cavities in trees
- hand pruner
- pole pruner
- disinfectant for sterilizing tools
- chisel and mallet
- battery-powered portable drill for assessing wood condition in trees
- increment borer for collecting wood samples from trees
- plastic bags for plant samples

- alcohol vials for insect samples
- hand lens (10× to 20×) for field inspections
- dissecting microscope for laboratory inspections
- insect net
- tape measure for measuring tree diameter
- binoculars for examining treetops

Figure 1.9. To collect complete information, it is important to consult with individuals who have knowledge of the site. Landscaping staff can provide valuable information regarding planting and maintenance history.

Figure 1.10. Diagnosticians should keep an open mind, because new or unusual problems may arise. The blue-stained sapwood in this stem of escallonia *(Escallonia rubra)* provided an indication that dieback may have been related to a janitorial cleaning solution being washed into the root zone.

Documentation

- camera
- notebook or tape recorder
- labels for samples (soil, water, or leaf)
- permanent marker
- pens and pencils

Step-by-Step Diagnostic Strategy

Approaching plant problem diagnosis in a systematic way helps you to be thorough and accurate, and it also allows you to identify corrective procedures that are specific to the problem.

1. Identify the Plant(s)

Identifying the plant is the essential first step in diagnosing a problem (fig. 1.11). Determine the genus, species, and cultivar (if appropriate). From the identity of the plant you can get an understanding of its origin, growth characteristics, and cultural requirements. Knowing the family that the plant belongs in may provide further information. If you cannot identify the plant, ask a colleague or submit a sample to a local botanical garden, arboretum, nursery, or UC Cooperative Extension office. You can easily misdiagnose a problem if you do not accurately identify the plant. Keep in mind that what appears to be a disorder may be normal for the species (see chapter 4). Refer to horticultural manuals for information regarding the cultural requirements of specific plants.

2. Identify the Symptoms

Examine the injured part(s) and list all symptoms. Be thorough and accurate. For a description of common symptoms, see chapter 3.

- Is the injured part chlorotic, necrotic, discolored, or distorted?
- Is the plant normal in size and development?

3. Inspect the Whole Plant

Examine all parts of the plant, not just the injured area (fig. 1.12).

- Is the injury confined to one part of the plant (e.g., leaves or flowers), or are other parts affected (e.g., multiple branches, bark, roots, fruit, flowers)? If injury symptoms are distributed uniformly throughout the canopy, the cause is often found in the root zone (fig. 1.13A). In some cases, however, an aboveground factor such as anthracnose can cause a uniform distribution of symptoms in the canopy (fig. 1.13B). If the injury is limited to discrete areas of the canopy, the cause is likely to be closely linked to the injured part (fig. 1.14).
- Is the plant growing, that is, are new leaves, flowers, or fruit being produced? Is the growth rate normal for the species or less than normal?
- Are there any signs of insects or disease organisms (e.g., droppings or cast skins of insects, boring dust from bark beetles, hyphae or mycellial plaques of fungi, foaming caused by bacteria, etc.)?

4. Inspect the Site

Examine the plant's environment and look for conditions that may contribute to injury (see fig. 1.12B). Evaluate soil physical and chemical properties (laboratory analysis of samples may be needed). Attempt to identify all factors that may have contributed to the symptoms identified in step 3.

- Is the terrain sloped?
- Does the soil appear to be well-drained?
- Are there soil layers or compacted zones?
- Is the area irrigated? If so, what is the irrigation schedule?
- Are there dry or wet areas?
- Is the site windy, hot, cold, shaded, exposed, or protected?
- Are there limitations to plant development due to infrastructure or hardscape (buildings, utilities, walkways, curbs, swimming pools, etc.)?
- Is there vegetation next to the problem plant(s)? If so, could it contribute to injury symptoms (e.g., by shading or by competing for water and minerals)?

5. Look for Patterns

- Are symptoms uniform throughout the plant, scattered, on one side only, or on old or young foliage (fig. 1.15)? Is one plant affected or are multiple plants affected?
- Are several species affected or only one species? For example, lack of soil oxygen caused by overwatering is likely to affect all of the plant species in a landscape, producing similar symptoms on all of the plants. On the other hand, plant diseases (biotic disorders) are more likely to affect a single species, or perhaps only those species in a specific family of plants.
- Do the symptoms form a pattern in the landscape? Are the symptoms related to topography (high spots, low spots, west-facing slopes)? Frost injury, for example, may affect susceptible plants more severely in low areas where cold air accumulates, rather than in higher, warmer areas of the landscape (see "Distinguishing between Biotic and Abiotic Disorders," above).
- Are symptoms related to wind patterns? Wind injury to shoots and leaves might be severe on the side of the plant facing the prevailing winds; the same injury might not be apparent on the downwind side of the plant.

6. Investigate the Plant Management History

Contact a person who has knowledge of the site and the plants in it. Collect as much background information as possible. Determine whether records were kept of landscape installation and maintenance. In many cases, you may have to rely on partial information. Be aware that, in some cases, inaccurate information may be given (e.g., where negligence may have contributed to plant injury).

- When was the site planted?
- How large (or how old) were the plants when they were planted?
- Have there been previous problems? If so, were they similar to the current problem?
- What are the irrigation, fertilization, soil

Traits of a Good Diagnostician

- Knowledge of plants, horticulture, soil, water, and pest management.
- Access to relevant resources.
- Knowledge of diagnostic techniques.
- An inquiring, open, logical mind.
- Patience and thoroughness.

Figure 1.11. Accurate plant identification is a key first step in diagnosis.

Figure 1.12. (A) Examine all plant parts during a diagnosis, not just the injured part. Although injury is evident on the leaves and branches of this tree, the trunk, root crown, and root zone should be inspected as well. (B) Carefully examine the plant environment for conditions that contribute to injury. In this planting, poor drainage is causing a soil aeration deficit and plant decline.

amendment, mulching, pruning, and pest management practices for the site?

- What is the water source? Has a water quality analysis been conducted?
- Have plant growth regulators or other chemicals been used?
- Have there been construction activities in the area?
- Have deicing salts been used on adjacent walkways or roadways?
- Have there been any other changes that may affect the landscape plants?

7. Synthesize Information

When you have finished collecting information about the plant and site, identify all potential causes of the problem.

- Be thorough and comprehensive.
- Identify the most likely causes.
- List reasons to support your diagnosis.

8. Test Likely Causes

Soil, water, or tissue samples are frequently needed for accurate diagnosis, particularly when nutrient deficiencies, specific ion toxicities, salts, or pH problems are suspected (for information on sample collection, see chapter 2). Submit samples to a reliable horticultural or agricultural laboratory. Use the results to create a list of likely causes.

- Once you have a list of likely causes, refer to chapter 5 for descriptions of specific abiotic disorders. If you suspect that the problem is caused by an organism (insect, fungus, bacteria, virus, nematode, or parasitic plant), refer to sources such as *Pests of Landscape Trees and Shrubs* (Dreistadt 1994).

- Determine whether your assessment is consistent with the description of the plant problem. If the description is consistent with the information you have collected, you have a good basis for making a reasonable diagnosis. If the descriptions are not consistent with information collected, you may need to reevaluate your list of likely causes. If you cannot arrive at a diagnosis, or if you are not confident of your diagnosis, seek assistance from knowledgeable specialists.

Figure 1.13. (A) If symptoms are expressed throughout the canopy, the cause of the problem may be found in the root zone. Dieback and chlorosis in the canopy of this Monterey pine *(Pinus radiata)* is uniformly distributed, and diagnosis should focus on the root zone. (B) In some cases, however, uniform dieback in the canopy may result from an aboveground cause. Here, anthracnose *(Stegophora ulmea)* has caused extensive defoliation of Chinese evergreen elm *(Ulmus parvifolia)*.

- Keep in mind that multiple causes are often involved. In addition, entirely new problems may arise, although this occurs more commonly with biotic disorders than abiotic disorders.
- In some cases, nothing can be done to correct a disorder. Although this is unfortunate from a plant management perspective, remedies that are unnecessary, inappropriate, or ineffective will be avoided.

Figure 1.14. If the injury is limited to a discrete part of the canopy, the cause is often closely linked to the injured part. Here, pitch canker *(Fusarium subglutinans)* infections on Monterey pine *(Pinus radiata)* branches have caused dieback of branch terminals.

Figure 1.15. Look for patterns and collect as much background information as possible. Finding injury on one side and at one level of all plants and knowing that this street was recently paved would provide a strong indication that injury was associated with heat released from paving equipment.

CHAPTER 2

Laboratory Analysis of Soil, Water, and Tissue Samples

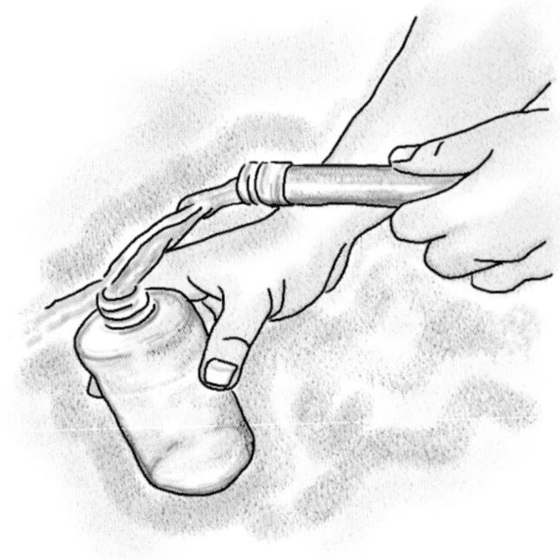

Figure 2.1. Laboratory analysis may be the only way to accurately diagnose certain disorders (e.g., salt stress, specific ion toxicities).

Figure 2.2. When collecting soil for laboratory analysis, gather samples from different locations around symptomatic and asymptomatic plants. Here, flags have been placed at sampling locations.

Diagnosing plant problems can be greatly aided by laboratory analysis of soil, water, and plant tissue samples (fig. 2.1). Disorders related to certain nutrient deficiencies, specific ion toxicities, salt stress, and sodic soil conditions can be diagnosed reliably only with the results of laboratory analysis. In some cases, testing may be the only way to confirm a diagnosis. For example, salt injury typically needs to be confirmed by an analysis of salt concentration in the soil; micronutrient deficiencies may require tissue analysis for confirmation. The following sections describe how to collect samples for laboratory submittal and which analyses are conducted for particular disorders.

Soil Testing

Collecting Samples

Determine where to sample

Collect soil samples within the root zone around symptomatic plants (fig. 2.2). If the surface soil has variation in color and texture, collect samples from each area that appears to be different. For plants in raised beds or planters, collect samples at several locations within the planter or bed. Remove surface vegetation, mulch, or other materials from each sampling site (fig. 2.3). Take samples from around asymptomatic plants as well as from around sympotmatic ones.

Determine how deep to sample

The location and depth of the root zone varies with the type of plant (trees, shrubs, groundcovers), species, and soil characteristics such as depth to hardpan, bulk density, layering, and moisture distribution (fig. 2.4). Dig a hole close to the sampling area to assess rooting depth. Typically, the top ½ inch (13 mm) of soil is discarded. To test for soil salinity, samples are typically taken at a depth of 1 to 6 inches (2.5 to 15 cm), but in some cases it may be necessary to sample deeper (see "Salinity" in chapter 5, p. 81). For herbicide analysis, sample in the 0- to 2-inch zone. Table 2.1 is a general guide to typical rooting and sampling depths.

Figure 2.3. Surface vegetation or mulch should be removed before collecting a soil sample in the root zone.

Determine how many samples to collect

Multiple samples are generally needed at each sampling location. This may require 10 to 15 subsamples for one composite sample at each location. Consult a soil laboratory for specific recommendations for particular sites. About 1 pint (about 0.5 l) of soil is needed from each sampling location.

Collect samples using a shovel, auger, soil sampling tube, or trowel (clean and rust-free). Pits dug with a backhoe are very useful for examining changes in soil color and texture, assessing root distribution, and for collecting deep samples. Air-dry samples and thoroughly mix them on a clean, dry surface before submitting for analysis (fig. 2.5). If available, use sampling bags provided by a soil laboratory. Otherwise, use

Figure 2.4. Collect samples from the root zone. Here, a soil sampling probe is used to collect soil in the root zone of a chlorotic sweetgum *(Liquidambar styraciflua)*.

Figure 2.5. Mix soil samples taken from one depth and location to make a composite sample. For soils that appear different (color and texture), collect separate composite samples.

TABLE 2.1. Soil sampling depths

Type of plant	Rooting depth		Depth at which to sample*	
	in	cm	in	cm
deep-rooted (trees and shrubs)	0–48	0–120	0–12 12–24 24–48	0–30 30–60 60–120
shallow-rooted (shrubs and groundcovers†)	0–18	0–46	0–6 6–12 12–18	0–15 15–30 30–46

Source: Harris, Clark, and Matheny 1999.
Notes:
*In raised beds or planters, sample the full depth of the container or bed.
†For groundcovers, sample at 0 to 6 inches (0–15 cm) and 6 to 18 inches (15–46 cm)

Figure 2.6. Place composite soil samples in a labeled sampling bag.

waterproof paper or plastic bags. Carefully label each sample on the outside of the bag (fig. 2.6).

Laboratory Analysis

Which analysis to request depends on which problems you suspect (see table 2.2). For example, to evaluate the level of salt in the soil, request an analysis of total soluble salts; for suspected sodic soil conditions, request salinity-alkalinity tests, which include the sodium adsorption ratio (SAR) or the exchangeable sodium percentage (ESP), which are determined from concentrations of sodium (Na), calcium (Ca), and magnesium (Mg).

Soil physical properties, including texture, bulk density, organic matter content and moisture release characteristics, depth, profile characteristics, and percolation, may need to be evaluated in order to diagnose abiotic disorders such as aeration deficits or water deficits (see table 2.3). Although soil texture can be assessed in the field, laboratory analysis is more reliable. Bulk density laboratory analysis requires a sample of specific volume (see Lichter and Costello 1994); contact a testing laboratory for instructions. Be careful not to compact samples when using a field core sampling tube. Organic matter content and moisture release characteristics should be determined by a laboratory. Depth, profile characteristics, and percolation can be determined in the field.

Results and Recommendations

Laboratories that specialize in horticultural or agricultural analyses can provide useful interpretations of results and recommendations to correct problems. Give the laboratory all pertinent information, including plant type, species, soil conditions, and relevant cultural practices, to help them customize their recommendations for your particular problem. For guidelines for interpretation of results, see table 5.6, p. 87.

Water Testing

The quality of irrigation water, whether groundwater, surface water, or recycled water, can be variable. Plants may be injured when poor-quality water is used. High concentrations of dissolved elements in irrigation water can lead to salt injury, specific ion toxicity, sodic soil conditions, and micronutrient deficiencies. Specific analyses must be conducted to diagnose particular disorders (table 2.3).

Collecting Samples

Samples must be representative of the water used to irrigate plants. Let the water run for a minute from the hose or irrigation system before collecting the sample. Collect samples in clean plastic bottles (fig. 2.7). Wash or rinse the bottles at least three times

TABLE 2.2. Soil chemical and physical analyses used to assess selected abiotic disorders

Suspected disorder	Soil analysis
salt injury	total soluble salt concentration: electrical conductivity (EC_e)
sodic soil	concentration of sodium (Na), calcium (Ca), and magnesium (Mg); calculate SAR (sodium adsorption ratio) or ESP (exchangeable sodium percentage)
micronutrient deficiency	soil pH (micronutrient deficiencies should be confirmed by tissue analysis or chelate application)
macronutrient deficiency	concentration of nitrogen (N), phosphorus (P), potassium (K), calcium (Ca^{++}); cation exchange capacity (CEC)
specific ion toxicity	concentration of chloride (Cl^-), boron (B), sodium (Na^+), and ammonium (NH_4^+),
water and aeration deficits caused by soil compaction	bulk density and texture
water and aeration problems caused by restrictive subsurface layers	depth to hardpan (field-test using auger or shovel) and percolation
aeration deficit caused by poor soil permeability	bulk density, texture, moisture release characteristics, percolation, and organic matter content
water stress caused by low water-holding capacity or poor infiltration	texture, moisture release characteristics, bulk density, and organic matter content

Figure 2.7. Collect water samples in a clean plastic container. Rinse the container 3 times with the water to be sampled.

Table 2.3. Water analyses used to assess selected abiotic disorders

Suspected disorder	Water analysis
salt injury	total soluble salts (EC_w)
specific ion toxicity	
root absorption	sodium (Na^+), chloride (Cl^-), boron (B)
foliar absorption	sodium (Na^+), chloride (Cl^-)
micronutrient deficiency caused by high pH	pH and bicarbonate concentration
water and aeration deficits caused by poor soil permeability	adjusted sodium adsorption ratio (adj. SAR), total soluble salts (EC_w)

with the water to be sampled. Laboratories usually need about 1 quart (about 1 liter) of water in the sample. The sample bottle should be filled completely. For certain analyses (check with the laboratory), the sample should be frozen or held below 40°F (4.4°C) until analyzed. Leave an air gap in the container if the sample is to be frozen.

Carefully label all samples. Imprecise labeling is a common problem. Include the date, location, source, sample number, and name of sampler. Keep notes regarding all sample label information.

Results and Recommendations

Guidelines for interpreting water quality analyses can be found in table 5.6, page 87; see also *Water Quality: Its Effects on Ornamental Plants* (Farnham, Ayers, and Hasek 1985). Interpretations of analyses should be made using these guidelines with consideration of specific plant and soil conditions. Laboratories that specialize in horticultural or agricultural water analyses may provide additional useful information.

Tissue Testing

It is best to select a new, fully expanded leaf for analysis (fig. 2.8). The entire leaf (blade and petiole) should be sent to the laboratory. For some plants, the petiole is used to analyze the status of chloride (Cl^-), nitrate-nitrogen (NO_3^--N), ammonium-nitrogen (NH_4^--N), and extractable potassium (K^+) and phosphorus (P). Leaf blades are used to analyze the status of potassium (K^+), calcium (Ca^{++}), magnesium (Mg^{++}), sodium (Na^+), iron (Fe^{++}), manganese (Mn), zinc (Zn), copper (Cu), boron (B), molybdenum (Mo), sulfate sulfur (SO_4^--S), and total nitrogen (N). See "Nutrient Deficiencies" in chapter 5 for more information regarding macronutrients and micronutrients.

Collect samples from both symptomatic and asymptomatic plants (fig. 2.9). Generally a minimum of 3.5 ounces (100 g) of fresh tissue is needed. On average, select 5 to 10 leaves from each plant. More leaves will be needed from species with small leaves (e.g., boxwood, *Buxus* spp.) than from species with larger leaves (e.g., southern magnolia, *Magnolia grandiflora*). Leaves should be equivalent in size, age, and condition. Place dry leaves in a clean paper bag that is clearly labeled and place the samples in a cool storage container for transport to the laboratory. Since plant nutrient status changes during the year, it may be necessary to take samples at different times during the growing season (early, middle, and late).

For guidelines for interpretation of test results, see table 5.6, p. 87.

Commercial Analytical Laboratories in California

Contact your local Cooperative Extension office or agricultural commissioner's office for a listing of analytical laboratories in your area. Select the appropriate lab for your needs:

- For soil, water, fertility, specific ion toxicity, and tissue analysis, choose a horticultural or agricultural laboratory.
- For bulk density and soil profile analysis, choose a soils or geotechnical laboratory.
- For herbicide and sludge analysis, choose an environmental testing laboratory.

Laboratories can recommend specific tests for the particular situation encountered, such as well water analysis or plant problem diagnosis.

Figure 2.8. Unless a laboratory recommends otherwise, collect new, fully expanded leaves from symptomatic plants. Select 5 to 10 leaves that are equivalent in size, age, and condition.

Figure 2.9. Select and submit leaves from asymptomatic plants as well as leaves showing injury (as separate samples). Here, nonchlorotic sweetgum *(Liquidambar styraciflua)* leaves are being collected to compare iron and manganese levels with chlorotic leaves.

CHAPTER 3

Common Injury Symptoms and Abiotic Causes

Symptoms are the external and internal reaction, response, or alteration of a plant as a result of a disease or injury. They are the signal to the diagnostician that something is wrong. Most diagnoses begin with an assessment of the type and pattern of symptoms in the plant and throughout the landscape. What may appear to be a symptom, however, may be a normal characteristic of the plant, such as premature fall coloration or peeling bark (see chapter 4).

Some symptoms are diagnostic, meaning that they are characteristic of a problem and lead directly to a diagnosis. Most often, however, symptoms can be caused by a variety of agents. The challenge for the diagnostician is to narrow down the possible causes and identify the specific problem. Quite often, biotic and abiotic agents cause similar symptoms. Observing the patterns of symptoms within the plant and throughout the landscape as well as noting species affected should help differentiate between biotic and abiotic agents. Although there are exceptions, biotic agents tend to be limited to one species or one plant, while abiotic agents may affect a wider range of species and multiple plants.

This chapter defines and illustrates terms used to describe symptoms, and it also identifies abiotic agents that can cause each symptom or pattern of symptoms. Beginning on p. 30, possible causes, patterns of symptoms, and methods for diagnosis are summarized in tabular form (table 3.1).

Figure 3.1. Wilt is an early symptom of water deficit.

Leaves

Wilting

Wilting is the loss of rigidity and drooping of leaves generally caused by insufficient water in the plant (fig. 3.1). Lack of water in the plant can be caused by:

- lack of available water in the soil (e.g., insufficient irrigation, drought, insufficient water-holding capacity in root area)
- inability of roots to take up adequate water (e.g., lack of soil aeration caused by too much water, mechanical injury to root system, exceptionally high demand for water)
- inability of xylem to move water from roots to shoots (e.g., girdling of stem)
- low temperatures that damage plant cells, causing them to leak water
- a variety of biotic agents (e.g., microorganisms, nematodes, root-feeding insects)

Leaf Necrosis

Leaf necrosis is death of all or part of a leaf. It can be caused by water deficit, salt and specific ion toxicity, herbicides, severe micronutrient deficiency, air pollution, high or low temperatures, excessive sun exposure, and other conditions. Many biotic agents also cause leaf necrosis.

The pattern of necrosis can be diagnostic. Patterns of necrosis on the leaf include:

- marginal: death of edges of leaves (fig. 3.2)
- blotch: death of large, irregularly shaped spots, or blots, on the leaf blade (fig. 3.3)
- spot: death of small, round areas on the leaf blade
- interveinal: death of leaf tissue between veins (fig. 3.4)
- entire leaf: death of entire leaf; if necrotic leaves are attached to the stem, death usually occurred quickly (fig. 3.5)

Note the pattern of necrosis on the entire plant. Damage on older leaves but not on younger leaves could indicate that the problem occurred earlier in the season, such as during a sudden springtime hot, windy period. It could also indicate a chronic problem, such as high soil salinity or specific ion toxicity (necrosis occurs when the ions reach toxic concentrations, and symptoms develop on older foliage before younger foliage). Necrosis on only one portion of the plant suggests a directional application of a toxic material to the foliage, such as salt spray or herbicide drift.

Note whether the necrosis is on one or two species within an area or on many plants throughout the area. If symptoms are limited to a few species, there may be variations in plant sensitivity to the abiotic

Figure 3.2. Marginal necrosis along edges of this eucalyptus (*Eucalyptus* sp.) leaf was caused by herbicide toxicity.

Figure 3.4. These interveinal necrotic symptoms on Lombardy poplar (*Populus nigra* 'Italica') were caused by a herbicide.

Figure 3.3. An irregularly shaped area of necrosis is called *blotch*, as illustrated here by anthracnose on London plane *(Platanus* × *acerifolia)*, a biotic problem.

agent, such as with salt toxicity. If all or most of the plants are affected, species tolerance may not be a factor, as could occur with the misapplication of a broad-spectrum herbicide.

Figure 3.5. Necrotic leaves attached to the stem indicates that death occurred quickly. This recently planted maidenhair tree *(Ginkgo biloba)* died within a few days from lack of water. The root ball was dry even though the surrounding turf was irrigated.

Chlorosis

Chlorosis is the yellowing of normally green tissue due to chlorophyll destruction or failure to form chlorophyll. Biotic agents, such as microorganisms, insects, and mites, as well as abiotic agents, can cause chlorosis. As with necrosis, observe the pattern of chlorosis. Chlorosis in young foliage may indicate deficiency of a nontranslocated element such as iron, manganese, or zinc. Chlorosis symptoms that develop first on older foliage may indicate a deficiency of elements that are translocated, such as nitrogen. Patterns of chlorosis include:

- general: overall yellowing of leaf blade
- interveinal: yellowing of tissue between the veins, with the veins remaining green (fig. 3.6)
- mottling or mosaic: irregular pattern of yellowing with shades of yellow to green (fig. 3.7)
- vein clearing: veinal tissue yellow or light-colored, with tissue between veins remaining green (fig. 3.8)

Figure 3.6. Interveinal chlorosis is yellowing of tissue between the veins.

Figure 3.7. Mottling, or mosaic, is an irregular pattern of varying shades of yellow to green.

Figure 3.8. Vein clearing, in which the veins are light-colored, can be a symptom of herbicide toxicity.

Figure 3.9. Silvering is a whitish appearance of a leaf, as illustrated in this creeping St. Johnswort *(Hypericum calycinum)* that was injured by thrips.

Figure 3.11. Leaves of this wax-leaf privet *(Ligustrum japonicum)* exhibit marginal chlorosis.

- stippling: small spots or flecks of light color within green leaf blade
- silvering: whitish-silver appearance of leaf surface due to damage to epidermis (fig. 3.9)
- bleaching: loss of color from leaf tissue (fig. 3.10)
- banding: alternating stripes of yellow and green on conifer needles
- marginal: yellowing along the leaf margin (fig. 3.11)

Ragged Leaves

Ragged leaves have been torn and mangled, usually due to wind, hail, or other abrasive action. Recent weather conditions are the key to diagnosis. Also notice the location of the symptoms relative to exposure. Biotic agents (feeding of insects or animals) may also create ragged leaves.

Malformed Leaves

Leaves can become malformed due to agents that act on the plant when the leaves are growing out of the bud (fig. 3.12). Abiotic causes typically are systemic herbicides;

Figure 3.10. Loss of color, or bleaching, in this white mulberry *(Morus alba)* was caused by a herbicide.

Figure 3.12. Leaves of this blackberry (*Rubus* sp.) are malformed due to toxicity from Roundup herbicide.

chlorosis, vein clearing, and/or necrosis usually accompany the symptom. Low temperatures during bud expansion can also cause malformation, as can a deficiency of zinc or manganese (see "Nutrient Deficiencies" in chapter 5). Malformed leaves without chlorosis or necrosis is generally caused by insects.

Defoliation

Defoliation is heavy loss of foliage at abnormal times of the year (fig. 3.13). Many agents can cause defoliation. Rapid defoliation may be caused by low temperatures or herbicides. Gradual defoliation may be caused by water deficit, soil aeration deficit, gas release in soil, or air pollution.

Figure 3.13. Defoliation is a loss of foliage, as illustrated in this London plane *(Platanus × acerifolia)* that was defoliated by a leaf disease (anthracnose).

Figure 3.14. Trees near this newly constructed home exhibit early fall color due to stress from root injury.

Early Fall Coloration and Premature Senescence

Premature senescence (earlier than normal onset of dormancy) in deciduous plants usually indicates chronic water stress. It is also a common symptom in plants whose root system has been injured (fig. 3.14).

Shoots

Wilting and Dieback

As with leaves, loss of rigidity and drooping of shoots is generally caused by insufficient water in the plant (fig. 3.15). Unless the water deficit is rectified, the symptoms usually progress to dieback of the shoot (fig. 3.16). Lack of water in a plant can be caused by

Figure 3.15. Wilting is a common symptom of water stress.

- lack of available water in the soil (e.g., insufficient irrigation, drought, insufficient water-holding capacity in root area)
- inability of roots to take up adequate water (e.g., lack of soil aeration caused by too much water, mechanical injury to root system, exceptionally high demand for water, high soil salinity)
- inability of xylem to move water from roots to shoots (e.g., girdling of stem)
- low temperatures that damage plant cells, causing them to leak water

Distortion

Shoot distortion is caused by an agent that acts on a plant as shoots elongate. Biotic agents are most commonly the cause. Possible abiotic agents include hormone-type herbicides; the herbicide would have been applied before shoot elongation occurred. Fasciation (abnormal flattening and coiling of the stem due to mutations in the cells) usually occurs on only one or a few branches and only on one plant (fig. 3.17). It is a genetic disorder that cannot be controlled.

Witches' Broom

Witches' broom is the proliferation of large numbers of shoots near the end of a branch (fig. 3.18). It is usually a symptom of a disease but may also be caused by genetic mutation. Possible abiotic causes include herbicide applications and disease.

Branches and Trunk

Sunken, Discolored Bark

Sunken, discolored bark is characteristic of sunburn (fig. 3.19). Symptoms are most severe where water deficits are also present.

Figure 3.16. Dieback of twigs and branches is a common symptom of prolonged water stress.

Figure 3.17. Flattening and coiling of this ash shoot (*Fraxinus* sp.) is fasciation caused by mutated cells.

Figure 3.18. The white witches' broom on this coast live oak *(Quercus agrifolia)* is caused by powdery mildew.

Figure 3.20. Woundwood is the new wood that has formed around this pruning wound.

Cankers caused by fungi or bacteria may create similar symptoms, but sunburn is confined to portions of the trunk or branches exposed to afternoon sun.

Woundwood Formation

Woundwood, sometimes called callus, is the lignified, partially differentiated tissue that develops around wounds (fig. 3.20). The presence of woundwood indicates that the bark has been injured by mechanical damage, sunburn, or freezing temperatures. Biotic agents also can cause injuries that stimulate woundwood production.

Figure 3.19. Sunburn causes cracked, sunken, discolored bark.

Galls

Galls are various abnormal growths that occur on plant parts. Some plants normally produce bud masses with swirling grain called lignotubers or burls (fig. 3.21). Galls may also result from biotic activity (bacteria, fungi, insects, nematodes, and mistletoe).

Bark Falling Off

Bark may peel and crack (exfoliate) when the plant dies or when vigor is low (fig. 3.22). It can be caused by sunburn, sprinkler water hitting the trunk frequently, water deficit, and lightning injury. Various biotic agents can cause bark exfoliation, or it may be normal for the species.

Bleeding and Gumming

Bleeding or gumming is the flow of sap from wounds or other injuries (fig. 3.23). It may be accompanied by a foul odor. Common abiotic causes are water deficit or mechanical injury. Bleeding may also be caused by biotic agents (e.g., canker fungi, bacteria), or it may be normal for the species.

Figure 3.21. Lignotubers, bud masses with swirling grain, are natural features of some trees.

Figure 3.22. The outer bark of this coast live oak *(Quercus agrifolia)* is cracking and exfoliating.

Figure 3.23. Gumming or sap flow through the bark is a common stress-induced symptom on many plants, including *Prunus* spp. and *Pinus* spp.

Swelling

Plant tissues do not actually swell, but areas may appear enlarged because of increased localized growth (fig. 3.24). Girdling from staking ties or circling roots are common causes. The stock below the graft union may also appear swollen. Localized enlarged areas on trunks and branches can indicate internal defects such as cracks (see Matheny and Clark 1994).

Splitting and Cracking

Herbicides, freezing temperatures, lightning, or mechanical failure may cause splitting of bark and cracks into the wood (fig. 3.25). Splits in the bark that do not extend into the wood can be caused by vigorous growth; these do not damage the plant.

Roots

Shriveled

Shriveled, collapsed, darkened roots indicate loss of turgor (fig. 3.26). This is a symptom of water deficits, high soil salinity, herbicide toxicity, or biotic disease.

Discolored

Normal root tissue is white to cream in color; discolored roots are black, gray, blue, or tan. Discoloration usually indicates loss of root function and root death. It may be caused by soil aeration deficits due to flooding, overirrigation, or placing a layer of fill soil over roots. Gas release in soil also causes root discoloration, and biotic root diseases cause similar symptoms. Some plants, such as the Australian tree fern *(Cyathea cooperi)*, normally have black roots.

Figure 3.24. The trunk of this pine is enlarged above the staking tie that restricted sap flow downward.

Figure 3.25. A crack through this Monterey cypress *(Cupressus macrocarpa)* branch is a result of mechanical failure due to wind.

Distorted

Abnormal root shapes such as club shapes or stunting are caused by soil-applied preemergent herbicides. The swollen appearance of roots or small galls can be caused by mycorrhizae or nitrogen-fixing bacteria that are beneficial to plants. Abnormal root growth also can be caused by plant parasitic nematodes (fig. 3.27).

Figure 3.26. This root ball has both healthy roots, which are whitish, and unhealthy roots, which are black.

Figure 3.27. Root distortion may be caused by biotic or abiotic agents. In this case, root knot nematode has caused atypical root growth.

Table 3.1. Summary of symptoms of abiotic disorders

Symptom	Possible cause	Pattern of symptoms	How to diagnose	Page
		Leaves		
Wilting	**A. Water deficit**			51–59
	1. Loss of roots			
	a. Mechanical injury (e.g., trenching through root system)	Sudden symptoms on all plants in affected area. Symptoms may be delayed until hot weather.	Determine history of site change; inspect for injury.	56
	b. Poor soil aeration			
	i. Flooding, poor drainage, high water table, overirrigation	Gradual development of symptoms affecting entire planting. Some species may be more sensitive than others.	Use soil probe or shovel to check moisture and color of soil. Blue or gray color and foul smell indicate anaerobic conditions.	60–65
	ii. Fill soil over roots, compacted soil	Gradual development of symptoms affecting entire planting. Some species may be more sensitive than others.	Excavate soil at base of tree to determine depth to original trunk flare.	64–66
	2. Water movement from roots to leaves interrupted due to mechanical damage to xylem, girdling	Entire or portion of plant may be affected. Not restricted to one species. Symptoms may develop gradually or suddenly, depending on extent of damage and transpirational demand.	Inspect plant for damage to bark. Check for girdling by staking wire. Check for girdling roots.	181–185
	3. Not enough water available for plant	Entire planting usually affected, especially drought-sensitive species. Symptoms develop suddenly. Plants with greatest sun and wind exposure show most severe symptoms.	Use soil probe to check soil moisture.	54
	a. Underirrigation	Entire planting usually affected, especially drought-sensitive species. Symptoms develop suddenly. Plants with greatest sun and wind exposure show most severe symptoms.	Check frequency and volume of water applied compared to plant need.	54–55
	b. Water not penetrating into soil	Entire planting usually affected, especially drought-sensitive species. Symptoms develop suddenly. Plants with greatest sun and wind exposure show most severe symptoms.	Use soil probe to check moisture. Run irrigation system and observe runoff and wetting patterns.	55
	i. Water running off slope	Entire planting usually affected, especially drought-sensitive species. Symptoms develop suddenly. Plants with greatest sun and wind exposure show most severe symptoms.	Use soil probe to check moisture. Run irrigation system and observe runoff and wetting patterns.	55
	ii. Surface hydrophobic or compacted	Entire planting usually affected, especially drought-sensitive species. Symptoms develop suddenly. Plants with greatest sun and wind exposure show most severe symptoms.	Evaluate soil volume and amount of available water relative to canopy size and water demand.	55
	c. Insufficient water-holding capacity in root area	Entire planting usually affected, especially drought-sensitive species. Symptoms develop suddenly. Plants with greatest sun and wind exposure show most severe symptoms.	Test water penetration through surface layer.	55

Symptom	Possible cause	Pattern of symptoms	How to diagnose	Page
		Leaves, cont.		
	i. Shallow soil	Entire planting usually affected, especially drought-sensitive species. Symptoms develop suddenly. Plants with greatest sun and wind exposure show most severe symptoms.	Investigate soil profile.	55
	ii. Very sandy soil	Entire planting usually affected, especially drought-sensitive species. Symptoms develop suddenly. Plants with greatest sun and wind exposure show most severe symptoms.	Determine soil texture.	55
	4. Exceptionally high demand for water (e.g., high temperature, wind, dry air)	Entire planting usually affected, especially drought-sensitive species. Symptoms develop suddenly. Plants with greatest sun and wind exposure show most severe symptoms. Plants usually recover overnight.	Determine weather conditions.	55
	B. Low temperature	Sudden, overall wilting on all cold-sensitive species in planting.	Determine recent low temperature and duration and plant sensitivity to chilling or freezing temperatures.	133–134
	C. Soil aeration deficit 1. Flooding, poor drainage, high water table, overirrigation	Gradual development of symptoms affecting entire planting. Some species may be more sensitive than others.	Use soil probe or shovel to check moisture and color of soil. Blue or gray color and foul smell indicate anaerobic conditions.	60–65
	2. Fill soil over roots, compacted soil	Gradual development of symptoms affecting entire planting. Some species may be more sensitive than others.	Excavate at base of tree to determine depth to original trunk flare.	64–66
	D. Gas release in soil (leaks, landfill)	Overall, gradual yellowing of all planting. Some species more sensitive than others. Symptoms progress to slow growth and death.	Check soil for foul odor and blue-gray or black color. Test soil atmosphere for gas. Cut into root and trunk tissue to check for blue or brown streaks in wood.	163–165
Leaf necrosis				
Marginal	**A. Water deficit** See "Leaves-Wilting-Water deficit."	See "Leaves-Wilting-Water deficit."	See "Leaves-Wilting-Water deficit."	52
	B. Toxicity 1. High soil salinity	Overall, gradual development of symptoms on entire planting. Necrosis most severe on older leaves. Tolerant species may not show damage.	Test soil and irrigation water for salinity, chloride. Determine history of salt application for snow and ice control.	81–82
	2. High boron content	Overall, gradual development of symptoms on entire planting. Necrosis most severe on older leaves. Tolerant species may not show damage. May appear as small necrotic spots along margin rather than entire margin.	Test soil, irrigation water, or plant tissue for boron.	82

Symptom	Possible cause	Pattern of symptoms	How to diagnose	Page
		Leaves, cont.		
Leaf necrosis	**B. Toxicity, cont.**			
	3. Herbicide (e.g., diuron, atrazine, dalapon, borates)	Sudden, overall symptoms affecting entire planting.	Determine possible herbicide category, and test.	188–192
	C. Severe iron deficiency	Overall, gradual development of symptoms on entire planting. Symptoms most severe on new growth. Tolerant species may not show damage.	Check for interveinal chlorosis.Test soil for carbonates, pH to determine if it is alkaline; test foliage for iron.	76–79, 118–119
Tip	**A. Air pollution**	Tip of young, expanding conifer needles turn red-brown, progressing to brown color.	Determine recent air quality and species sensitivity.	167–168
	B. High soil salinity	Overall, gradual development of symptoms on entire planting. Necrosis most severe on older leaves. Tolerant species may not show damage.	Test soil for salinity, chloride. Determine history of salt application for snow and ice control.	81–82
	C. High boron content	Overall, gradual development of symptoms on entire planting. Necrosis most severe on older leaves. Tolerant species may not show damage. May appear as small necrotic spots along margin rather than entire margin.	Test soil, irrigation water, and/or plant tissue for boron.	82
Blotch	**A. Water deficit** See "Leaves-Wilting-Water deficit."	See "Leaves-Wilting-Water deficit."	See "Leaves-Wilting-Water deficit."	52
	B. Too much sun for species	Sudden symptoms on outside of canopy affecting sun-sensitive plants. May start with yellowing on tips and margins of leaves.	Determine exposure, reflected heat, and plant tolerances. More severe when coupled with low soil moisture. May appear following pruning, removal of adjacent plant, or increased exposure.	142
	C. High temperature	Entire planting usually affected, especially heat-sensitive species. Symptoms develop suddenly. Plants with greatest sun and wind exposure, sow soil moisture show most severe symptoms.	Determine weather conditions.	139, 142
Spot	**A. Herbicide** (e.g., paraquat, bentazon, diphenyl-ethers, oxadiazon, diquat, fluazifop, cyclohexenone)	Sudden symptoms with greatest damage on outer leaves present at the time of exposure. Spots usually uniform in size, color, and distribution, with sharp margin.	Determine what herbicides have been used nearby. Submit leaf samples to laboratory for testing.	188–192
	B. Air pollution	Sudden or gradual symptom development, depending on level and type of pollutants. Affects entire planting of sensitive species.	Determine recent air quality and species sensitivity.	167–168
Interveinal	**A. Air pollution**	Sudden or gradual symptom development, depending on level and type of pollutants. Affects entire planting of sensitive species.	Determine recent air quality and species sensitivity.	167–168

Symptom	Possible cause	Pattern of symptoms	How to diagnose	Page
		Leaves, cont.		
	B. Herbicide (e.g., simazine, diuron, bromacil, atrazine, terbacil)	Sudden symptom development affecting entire planting.	Determine what herbicides have been used nearby. Submit leaf samples to laboratory for testing.	188–192
	C. Severe manganese deficiency	Gradual symptoms development on old and new foliage. Symptoms most severe on species sensitive to alkaline soil.	Submit soil sample to laboratory to test for pH, calcium carbonate. Submit leaf samples to laboratory for manganese testing.	78–79, 118–119
Entire leaf	**A. Water deficit** See "Leaves-Wilting-Water deficit."	See "Leaves-Wilting-Water deficit."	See "Leaves-Wilting-Water deficit."	52
	B. Herbicide (e.g., weed oil, diquat, paraquat, oxyfluorfen)	Sudden symptom development affecting entire planting.	Determine what chemicals have been applied. Submit soil and/or leaf samples for testing.	188–192
	C. Low temperature: Frost	Sudden symptom development affecting entire planting, limited to cold-sensitive species. Damage may be more severe in low-lying areas.	Determine recent low temperature and duration and plant sensitivity to chilling or freezing temperatures.	133–134
	D. Low temperature: Winter injury	Common in evergreens in late winter or early spring following winds. Most severe on side of plant exposed to wind.	Determine if soil is frozen.	133–134
	E. Air pollution	Dead tissue extending all the way through leaf, white to dark orange-red. Larger veins may remain green. Affects entire planting of sensitive species.	Determine recent air quality and species sensitivity.	167–168
Chlorosis	**Normal for species**	Variation in leaf color is often a normal characteristic of a species or cultivar.		42–44
General	**A. Nitrogen deficiency**	Gradual symptom development affecting entire planting. Symptoms most severe on older foliage first.	Determine when fertilizer last applied. Test foliage for nitrogen concentration.	71–72
	B. Damaged root system 1. Saturated soil	Overall, gradual development of symptoms affecting entire planting.	Use soil probe or shovel to check moisture and color of soil. Blue or gray color and foul smell indicate anaerobic conditions.	60–65
	2. Fill soil over roots	Gradual development of symptoms affecting entire planting.	Excavate soil at base of plant to determine depth to original trunk flare.	66
	3. Mechanical damage	A portion or the entire plant may be affected. Chlorosis symptoms usually gradual, but plant may wilt suddenly during hot weather.	Determine history of site change; inspect for injury.	56, 181–185
	C. Moderate soil salinity or sodic soil	Overall gradual symptom development affecting salt-sensitive plants in area.	Submit soil sample to laboratory to test for salinity and SAR.	81–82

Symptom	Possible cause	Pattern of symptoms	How to diagnose	Page
		Leaves, cont.		
Chlorosis, General, cont.				
	D. Girdled trunk or roots			
	1. Staking ties embedded in trunk	Gradual symptom development on individual plant.	Examine staking ties for damage to trunk.	181
	2. Sunburned trunk	Damage only to portions of bark exposed to afternoon sun. Most severe on thin-barked trees. Damage worse when soil moisture low.	Examine sides of trunk exposed to afternoon sun for dead and damaged bark.	139, 144
	3. Mechanical damage	Entire plant usually affected. Not restricted to one species. Symptoms may develop gradually or suddenly, depending on extent of damage and transpirational demand.	Check root crown, trunk, and major branches for wounds.	181–185
	E. Gas release in soil (leaks, landfill)	Overall, gradual yellowing of all planting. Some species more sensitive than others. Symptoms progress to slow growth and death.	Check soil for foul odor and blue-gray or black color. Test soil atmosphere for gas. Cut into root and trunk tissue to check for blue or brown streaks in wood.	163–165
	F. Air pollution	Sudden or gradual symptom development, depending on level and type of pollutants. Affects entire planting of sensitive species.	Determine recent air quality and species sensitivity.	167–168
Interveinal	**A. Micronutrient deficiency**			
	1. Iron deficiency	Overall, gradual development of symptoms on entire planting. Symptoms most severe on young growth. New growth may be bleached and in severe cases develop necrotic areas. Species tolerant of alkaline soil may not show damage.	Submit soil sample to laboratory to test for pH, calcium carbonate. Submit leaf samples to laboratory for testing. Symptoms may be temporary if soils are cold or wet.	76, 77, 118–119
	2. Manganese deficiency	Overall, gradual development of symptoms on entire planting. Band of green tissue around veins wider than iron deficiency. Symptoms on both old and new growth. Leaves may develop necrotic spots. Species tolerant of alkaline soil may not show damage.	Submit soil sample to laboratory to test for pH, calcium carbonate. Submit leaf samples to laboratory to test for manganese concentration.	78-79, 118–119
	B. Specific ion toxicity: chloride, boron, sodium, or ammonium, in incipient stages	Overall, gradual development of symptoms on sensitive plants. Older leaves affected first. Damage progresses to necrosis.	Submit soil, water, and/or leaf samples to laboratory to test for chloride, boron, and sodium.	82–83
	C. Herbicide (e.g., triazines, atrazine, simazine)	Greatest chlorosis along tips, margins, and veins on leaves, beginning in older foliage. Entire planting affected.	These materials are soil-applied. Submit soil samples to laboratory for testing.	188–192
Mottling/ mosaic	**A. Zinc deficiency**	Young foliage develops symptoms first. Leaves may be abnormally small and necrotic, with short internodes (rosettes). More than one species may show symptoms.	Submit soil sample to laboratory to test for pH, calcium carbonate. Submit leaf samples to laboratory to test for zinc concentration.	79–80, 118–119

Symptom	Possible cause	Pattern of symptoms	How to diagnose	Page
		Leaves, cont.		
	B. Herbicide (e.g., uracils, bromacil, terbacil)	Sudden symptoms on entire planting exposed to chemical.	Submit soil samples to laboratory for testing.	188–192
Vein clearing	Herbicide (e.g., substituted ureas, diuron, monuron, neburon)	Sudden symptom development on entire planting. Interveinal tissue remains green. May affect a portion or all of plant.	Determine what chemicals have been applied. Submit soil samples and to laboratory for testing.	188–192
Stippling	**A. Air pollution**	Sudden or gradual symptom development, depending on level and type of pollutants. Affects entire planting.	Determine recent air quality and species sensitivity.	167–168
	B. Release of gas into atmosphere	Sudden symptom development on entire planting area exposed. Injury symptoms greatest nearest gas release location.	Look for potential sources of phytotoxic gas (e.g., vents, chimneys, smokestacks, industrial equipment).	163–165
Silvering/ bronzing	**A. Air pollution**	Sudden or gradual symptom development, depending on level and type of pollutants. Affects entire planting of sensitive species.	Determine recent air quality and species sensitivity.	167–168
	B. Atmospheric gas release	Sudden symptom development on entire planting area exposed. Injury symptoms greatest nearest gas release location.	Look for potential sources of phytotoxic gas (e.g., vents, chimneys, smokestacks, industrial equipment).	163–165
Bleaching	**A. Herbicide** (e.g., amitrole, norflurazon, clomazone)	Sudden symptoms on new leaves, particularly at the tips. Entire planting exposed to chemical is affected.	Determine what chemicals have been applied. Submit soil samples to laboratory for testing.	188–192
	B. Iron deficiency	Symptoms most severe on young growth. Necrotic areas may develop. Species tolerant of alkaline soil may not show damage.	Submit soil sample to laboratory to test for pH, calcium carbonate. Submit leaf samples to laboratory for testing. Symptoms may be temporary if soils are cold or wet.	76–77
	C. Atmospheric gas release	Sudden symptom development on entire planting area exposed. Injury symptoms greatest nearest gas release location.	Look for potential sources of phytotoxic gas (e.g., vents, chimneys, smokestacks, industrial equipment).	163–165
Banding (conifers)	**Air pollution**	Sudden or gradual symptom development, depending on level and type of pollutants. Affects entire planting. Develops first on semimature needles.	Determine recent air quality and species sensitivity.	167–168
Early fall coloration	**A. Water deficit**	Gradual symptoms on deciduous species sensitive to drought, root cutting, or other factors reducing water availability.	Use soil probe to check soil moisture. Investigate previous incidences of root cutting, soil compaction, and construction activities.	53

Symptom	Possible cause	Pattern of symptoms	How to diagnose	Page
		Leaves, cont.		
Early fall coloration, cont.				
	B. Nitrogen deficiency	Gradual symptoms on susceptible species.	Determine history of fertilization. Submit tissue samples to laboratory to test for nitrogen in spring or summer.	71–73
Ragged leaves	**Weather damage** (e.g., wind, hail)	Sudden symptoms on all plants exposed to damaging agent.	Determine recent weather conditions.	158–160, 175–176
Malformed leaves	**A. Normal for species**	Some plants normally have unusally formed leaves.		42–44
	B. Herbicide (e.g., glyphosate, sulfonylureas, imidazolinones)	New growth develops with deformation symptoms as well as tip chlorosis and dieback. All plants within treated area affected.	Determine what chemicals have been applied. Glyphosate injury may not be seen until spring following fall application. For soil-applied chemicals, submit soil samples to laboratory for testing.	188–192
Defoliation/ premature senesence	**A. Normal for species**	Some evergreens normally have heavy drop of oldest leaves at various times of year.		40–42
	B. Water deficit			
	See "Leaves-Wilting-Water deficit."	See "Leaves-Wilting-Water deficit."	See "Leaves-Wilting-Water deficit."	52–53
	C. Low temperature: Frost	Sudden symptom development affecting entire planting.	Determine recent temperatures and plant hardiness.	133–135
	D. Herbicide (e.g., bromacil)	Sudden symptoms on entire planting in treated area. Defoliation may be preceded by marginal necrosis. Older foliage affected first.	Determine what chemicals have been applied. Submit soil samples to laboratory for testing.	188–192
	E. Soil aeration deficit			
	1. Flooding, poor drainage, high water table, overirrigation	Gradual development of symptoms affecting entire planting. Some species may be more sensitive than others.	Use soil probe or shovel to check moisture and color of soil. Blue or gray color and foul smell indicate anaerobic conditions.	60–65
	2. Fill soil over roots, compacted soil	Gradual development of symptoms affecting entire planting. Some species may be more sensitive than others.	Excavate at base of tree to determine depth to original trunk flare.	66
	3. Sodic soil	Entire planting usually affected.	Test soil for SAR.	163–165
	F. Gas release in soil (leaks, landfill)	Overall, gradual wilting and dieback of all plants in affected area. Some species more sensitive than others.	Check soil for foul odor and blue-gray or black color. Test soil atmosphere for gas. Cut into root and trunk tissue to check for blue or brown streaks in wood.	82
	G. Air pollution (chronic)	Affects entire planting.	Determine recent air quality and species sensitivity.	167–178

Symptom	Possible cause	Pattern of symptoms	How to diagnose	Page
		Shoots		
Wilting and/or dieback	**A. Water deficit** See "Leaves-Wilting-Water deficit."	See "Leaves-Wilting-Water deficit."	See "Leaves-Wilting-Water deficit."	52
	B. Low temperature: Frost	Sudden, overall symptoms on young shoots of all cold-sensitive species in planting.	Determine recent temperatures.	133–134
	C. Gas release in soil (leaks, landfill)	Overall, gradual wilting and dieback of young shoots of all plants in affected area. Some species more sensitive than others.	Check soil for foul odor and blue-gray or black color. Test soil atmosphere for gas. Cut into root and trunk tissue to check for blue or brown streaks in wood.	163–165
	D. Soil aeration deficit 1. Flooding, poor drainage, high water table, overirrigation	Gradual development of symptoms affecting entire planting. Some species may be more sensitive than others.	Use soil probe or shovel to check moisture and color of soil. Blue or gray color and foul smell indicate anaerobic conditions.	60–65
	2. Fill soil over roots, compacted soil	Gradual development of symptoms affecting entire planting. Some species may be more sensitive than others.	Excavate soil at base of plant to determine depth to original trunk flare.	63–66
Distortion	**A. Herbicide** (e.g., thiocarbamates, 2,4-D, dicamba, glyphosate, dalapon)	Symptoms on new growth throughout planting.	Thiocarbamates are applied to soil; submit soil samples to laboratory for testing. 2,4-D and dicamba may be in fertilizers and weed control mixtures applied to turf.	188–192
	B. Fasciation	Symptoms usually on only one or a few branches, and only one plant.	Abnormal flattening and coiling of stem due to mutation in cell.	36, 49
Witches' broom	**A. Cell mutation**	Symptoms usually on only one or a few branches, and only one plant.	Eliminate other causes.	49
	B. Herbicide (e.g., glyphosate)	New growth develops with deformation of young leaves and proliferation of new buds without elongation. All plants within treated area affected.	Determine what chemicals have been applied. Glyphosate injury may not be seen until spring following fall application.	188–192
		Branches, Trunk		
Sunken, discolored bark	**Sunburn or sunscald**	Symptoms on portions of trunk exposed to afternoon sun.	Check for cracked and loose bark only on the southwest side of trunk; location is generally diagnostic.	139–140, 144–145
Wound-wood formation	**A. Mechanical damage or failure**	Isolated area at point of impact or injury. Gradual development.	Inspect plant for damage.	181–185
	B. Sunburn or sunscald	Symptoms on portions of trunk exposed to afternoon sun.	Check for cracked and loose bark only on the southwest side of trunk; location is generally diagnostic.	139–140, 144–145

Symptom	Possible cause	Pattern of symptoms	How to diagnose	Page
		Branches, Trunk, cont.		
Woundwood formation, cont.				
	C. Freezing temperatures	After repeated splitting, woundwood ridges form along edges of crack and wood decay may begin. Trees 6 to 18 inches (15 to 45 cm) in diameter more likely to be affected than smaller or larger trees.	Determine winter low temperatures and examine tree for associated defects.	133–135, 144–145
Galls	**Normal for species**	Bud masses and swirling grain (burls) are normal for some plants.		45–48
Bark falling off	**A. Normal for species**	Peeling and cracking bark as it matures is a normal characteristic of some trees.		44–45
	B. Sunburn or sunscald	Located on portions of trunk exposed to afternoon sun.	Check for cracked and loose bark only on the southwest side of trunk; location is generally diagnostic.	139–140, 144–145
	C. Sprinkler water hitting trunk	Damaged area corresponds to sprinkler pattern. Most common on trees with thick corky bark and palms.	Determine sprinkler pattern.	37
	D. Low vigor	Gradual, overall decline in plant health.	Tree decline may lead to bark exfoliation.	37
	E. Water deficit	See "Leaves-Wilting-Water deficit." Most severe on portions of trunk exposed to afternoon sun.	See "Leaves-Wilting-Water deficit"	53
	F. Lightning injury	Occurs most commonly on lone, tall trees, windward edges of stands, and dominant trees in stands. Damaged branches may die.	Look for strip of bark removed from top to bottom of tree, with a rough groove that follows the grain of wood.	172–174
Bleeding, gumming	**A. Normal for species**	Exudation of sap is a normal characertistic in some species.		27
	B. Water deficit	See "Leaves-Wilting-Water deficit." Common on stone fruits (peach, cherry, etc.).	See "Leaves-Wilting-Water deficit."	53
	C. Mechanical injury (e.g., pruning cuts)	No pattern.	Look for injured tissue.	181–185
	D. Poor soil aeration	Symptoms may be limited to species prone to gumming.	Check soil moisture content, color, and odor.	61–62
Swelling	**A. Graft union**	In grafted trees, diameter of stock is much larger than that of scion at the graft union.	Not considered a problem.	186
	B. Girdling	Swelling of tissue above girdle.		177–180
	1. Staking tie left too long		Examine trunk for damage by staking ties.	181–185
	2. Girdling root		Excavate soil to examine root crown and determine if girdling root present.	177–179
Splitting, cracking	**A. Herbicide** (e.g., phenoxy compounds)	Susceptible woody plants exposed to chemical. Usually accompanied by foliar and shoot distortion.	Determine what chemicals have been applied. 2,4-D and dicamba may be in fertilizers and weed control mixtures applied to turf.	188-192

Symptom	Possible cause	Pattern of symptoms	How to diagnose	Page
		Branches, Trunk, cont.		
	B. Freezing temperatures	Longitudinal cracks usually associated with wounds and branch stubs. After repeated splitting, woundwood ridges form along edges of crack and wood decay may begin. Trees 6 to 18 inches (15 to 45 cm) in diameter more likely to be affected than smaller or larger trees.	Determine winter low temperatures and examine tree for associated defects.	133–135
	C. Growth cracks	Occurs in areas of vigorous growth, especially in species with thick bark. Indication of vigorous growth.	Crack in corky bark does not penetrate to inner bark and wood.	45
	D. Mechanical failure	Most common on individual branches or trunks with excessive weight, especially when exposed to strong wind, ice, snow or rain.	Crack extends into wood. Wood fibers pulled apart.	39
	E. Lightning injury	Split following grain of wood from point of strike to ground.	Uncommon in urban areas of California and in areas not prone to lightning storms.	172–174
		Roots		
Shriveled	**A. Water deficit**	See "Leaves-Wilting-Water deficit."	See "Leaves-Wilting-Water deficit."	51–59
	B. High soil salinity	Overall, gradual development of symptoms on entire planting. Necrosis most severe on older leaves. Tolerant species may not show damage.	Test soil for salinity, chloride. Determine history of salt application for snow or ice control.	81–82
	C. Herbicides (e.g., dinitroanilines, amides, thiocarbamates)	Overall, gradual development of symptoms on entire treated area. Plants usually stunted, may show general or interveinal chlorosis.	Submit soil samples to laboratory to test for salinity.	190
Discolored	**A. Soil aeration deficit**			
	1. Flooding, poor drainage, high water table, overirrigation	Gradual development of symptoms affecting entire planting. Some species may be more sensitive than others.	Use soil probe or shovel to check moisture and color of soil. Blue or gray color and foul smell indicate anaerobic conditions.	60
	2. Fill soil over roots, compacted soil	Gradual development of symptoms affecting entire planting. Some species may be more sensitive than others.	Excavate at base of tree to determine depth to original trunk flare.	60, 66
	B. Gas release in soil (leaks, landfill)	Overall, gradual wilting and dieback of all plants in affected area. Some species more sensitive than others.	Check soil for foul odor and blue-gray or black color. Test soil atmosphere for gas. Cut into root and trunk tissue to check for blue or brown streaks in wood.	163–165
Distorted	**A. Herbicide** (e.g., trifluralin, oryzalin)	Overall, gradual development of symptoms on entire treated area. Plants usually stunted, may show general or interveinal chlorosis.	Determine what chemicals have been applied; submit soil samples to laboratory for testing.	190
	B. Normal for species	Some species form associations with nitrogen-fixing bacteria that produce small growths on roots.		47–48

Source: Harris, Clark, and Matheny 1999.

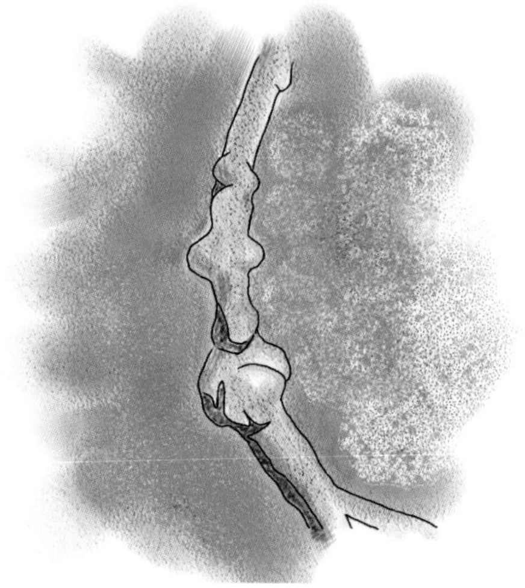

CHAPTER 4

Plant Traits That Resemble Abiotic Disorders

A number of common landscape tree and shrub species have physical or physiological traits that may resemble abiotic or biotic disorders. Although these may appear to be problems, they are simply natural growth characteristics. It is important to identify a plant correctly and know its "normal" growth characteristics to avoid mistaking a natural characteristic for a problem. This chapter describes natural traits that may be found on or associated with many landscape plants.

Natural Leaf Shedding (Leaf Senescence)

Natural leaf shedding is often mistaken for a disorder. Trees and shrubs renew their foliage each year: old leaves fall off, and new leaves are formed. The leaves of deciduous species are retained for several months, while leaves of evergreens live from one to several years. The amount of leaf shedding in evergreen species depends on the species and is often influenced by environmental factors; for instance, heavy leaf shedding often follows a cold winter.

In broadleaf evergreens, old leaves throughout the canopy turn yellow to straw-colored, then drop. In some cases, as much as one-quarter to one-third of the foliage becomes chlorotic and drops within a month. The plant may be perceived as declining during this period. However, the terminal buds and usually the one- or two-year-old leaves remain healthy and grow normally. Certain broadleaf evergreen species shed old leaves in early spring, just before or while new leaves begin to grow. Examples include cork oak (*Quercus suber*), holly oak (*Quercus ilex*), coast live oak (*Quercus agrifolia*), interior live oak (*Quercus wislizenii*), camphor tree (*Cinnamomum camphora*), xylosma (*Xylosma congestum*), and privet (*Ligustrum* spp.) (fig. 4.1). Other species, such as southern magnolia (*Magnolia grandiflora*), carob (*Ceratonia siliqua*), and eucalyptus (*Eucalyptus* spp.), shed some foliage throughout spring and summer.

In evergreen conifers such as pine (*Pinus* spp.), cedar (*Cedrus* spp.), and coast redwood (*Sequoia sempervirens*), old foliage throughout the canopy turns yellow to brown, then drops. Foliage is normally shed in fall, although some may drop in spring following a cold, wet winter or when drought-stressed. In pines, needles remain together in fascicles of 2 or 5 when they drop. As with broadleaf evergreens, buds and new foliage are retained (fig. 4.2).

Certain conifers are deciduous, and their natural leaf drop is often confused with a symptom of disorder. In dawn redwood (*Metasequoia glyptostroboides*), larch (*Larix* spp.), and bald cypress (*Taxodium distichum*), for example, foliage is naturally shed in fall along with leaves of other deciduous species. The leaves turn yellow to reddish brown before falling (figs. 4.3–4).

Other species are referred to as being summer- or drought-deciduous, dropping leaves in order to reduce transpirational water loss during hot, dry periods. For example, California buckeye (*Aesculus californica*) and box elder (*Acer negundo*) shed leaves in summer (July–August) to survive conditions of low soil moisture. Leaves turn

Figure 4.1. In spring, natural leaf shedding in cork oak *(Quercus suber)* can be mistaken as a disorder (A). Only older leaves are being shed, and new leaves are beginning to grow (B). After approximately 4 weeks, the same tree is covered with new leaves (C).

brown rapidly but drop over a period of several weeks (fig. 4.5).

Normal leaf shedding in evergreens may be confused with symptoms caused by root and crown diseases. Normal fall senescence or dormancy of deciduous conifers and normal summer senescence or dormancy of drought-deciduous species may also resemble symptoms

Figure 4.2. Fall leaf shedding in coast redwood *(Sequoia sempervirens)* may appear to be a biotic or abiotic disorder (A). Only older leaves are being shed, while terminal foliage is green and healthy (B).

caused by root disorders, severe root injury, or root and crown rot diseases.

No remedies are needed for natural leaf shedding. Supplemental irrigation may prevent some drought-deciduous species from dropping leaves in midsummer.

Leaf Retention in Deciduous Species

Leaves showing fall color can be retained well into the winter on certain deciduous species. Most notably, pin oak *(Quercus palustris)* may not drop leaves until midwinter in some locations. Typically, retained leaves are russet brown, making the trees appear to be severely water-stressed or dead (fig. 4.6).

Plant Senescence

A few plant species decline and die after blooming and setting seed. Examples are agave (*Agave* spp.) and several species of bamboo (*Phyllostachys* spp.). With bamboo, only a portion of the plant may bloom, set seed, and die. Such dieback is a natural characteristic of the plant species and is not related to diseases or disorders (fig. 4.7).

Natural Leaf Colors: Variegated Leaves, Red Spring Foliage

Some plants normally have leaves that are striped, banded, spotted, or blotched with yellow, white, red, and other colors in combination with green. The colored areas do not gradually blend into the green area of the leaf but are usually sharply defined with

Figure 4.3. Fall leaf drop in bald cypress *(Taxodium distichum),* a natural characteristic of this deciduous conifer. Because conifers are usually evergreen, this is often confused with dieback caused by root diseases or other problems.

Figure 4.5. California buckeye *(Aesculus californica),* a drought-deciduous tree, defoliated in June in Woodside, California. This is a natural characteristic of the species that can be delayed, but not prevented, by applying water during the summer.

Figure 4.4. Fall leaf drop in larch (*Larix* spp.), a deciduous conifer.

Figure 4.6. Deciduous species that retain leaves during the winter may appear to be severely water-stressed or dead. Here, russet leaves on pin oak *(Quercus palustris)* in January make it appear to have a serious problem (A). Young pin oaks look dead during the winter (B).

Figure 4.7. Golden bamboo *(Phyllostachys aurea)* will die after having bloomed and set seed.

definite boundaries. This natural characteristic is called *variegation*. Species with distinctive variegations include flowering maple *(Abutilon megapotamicum)*, box elder *(Acer negundo)*, evergreen euonymus *(Euonymus japonica)*, geranium (*Pelargonium* spp.), English holly *(Ilex aquifolium)*, ivy (*Hedera* spp.), holly-leaf osmanthus *(Osmanthus heterophyllus* 'Variegatus'), Japanese pittosporum (*Pittosporum tobira* 'Variegata'), true myrtle (*Myrtus communis* 'Variegata'), and laurustinus (*Viburnum tinus* 'Variegatum') (fig. 4.8).

Leaf variegations may be confused with chlorosis caused by micronutrient deficiencies or pesticide phytotoxicities. However, normal variegations occur on leaves of all ages, while variegation caused by micronutrient deficiencies (iron, manganese) occurs only on new leaves. Chlorosis caused by pesticide phytotoxicity is generally accompanied by other symptoms, such as necrosis and dieback.

Figure 4.8. Variegated leaves of true myrtle (*Myrtus communis* 'Variegata') (A) and evergreen euonymus *(Euonymus japonica)* (B). Natural leaf variegation sometimes is confused with nutrient deficiency.

Figure 4.9. Red-colored new leaves of Chinese photinia *(Photinia serrulata)* (A) and rose (*Rosa* sp.) (B). These leaves will become green as the season progresses.

The new leaves of some species, including Chinese photinia *(Photinia serrulata)*, Japanese maple *(Acer palmatum)*, rose (*Rosa* spp.), and apricot *(Prunus armeniaca)*, are red or copper-colored, especially in spring. These leaves eventually turn green as the season progresses (fig. 4.9). Similarly, the leaves of some species, including Japanese cryptomeria *(Cryptomeria japonica)* and heavenly bamboo (*Nandina* spp.), turn various shades of red during the winter months in response to low temperatures (fig. 4.10). The distinctive colors of these leaves may be confused with nutrient deficiencies or chemical phytotoxicities.

Natural Bark Characteristics: Corky Bark, Peeling or Flaking Bark, Bark Cracks

Some landscape species have distinctive natural bark characteristics. Corky "wings" on the twigs of American sweetgum *(Liquidambar styraciflua)* and elm (*Ulmus* spp.) create a roughened appearance that may be mistaken for disease or insect infestation (fig. 4.11).

The trunks and branches of several species, including eucalyptus (*Eucalyptus* spp.), river birch *(Betula nigra)*, Russian olive *(Elaeagnus angustifolia)*, sycamore

(*Platanus* spp.), and crape myrtle (*Lagerstroemia indica*), have bark that peels or falls in flakes (exfoliating bark) (fig. 4.12). Newly formed bark beneath the peeling bark is smooth and tightly attached to the wood. Although exfoliating bark is natural in some species, root diseases and mechanical bark injuries may cause loose or peeling bark in many other species.

Deep longitudinal cracks or furrows (also called growth cracks, tension cracks, or expansion cracks) may occur in the bark of many tree species as the tree grows in diameter and the bark does not expand as rapidly as the underlying wood (fig. 4.13). The cracks usually become less obvious as the bark thickens in late summer, and they eventually close. Bark cracks often occur in fast-growing tree species.

Cracked or peeling bark that is similar to natural bark cracks may be caused by root diseases, sunburn, drought, or mechanical bark injuries that destroy cambium tissue. In such cases, the underlying wood becomes exposed as the bark dies and sloughs off.

Figure 4.10. In response to low temperatures in winter, foliage of Japanese cryptomeria (*Cryptomeria japonica* 'Elegans') turns plum-colored.

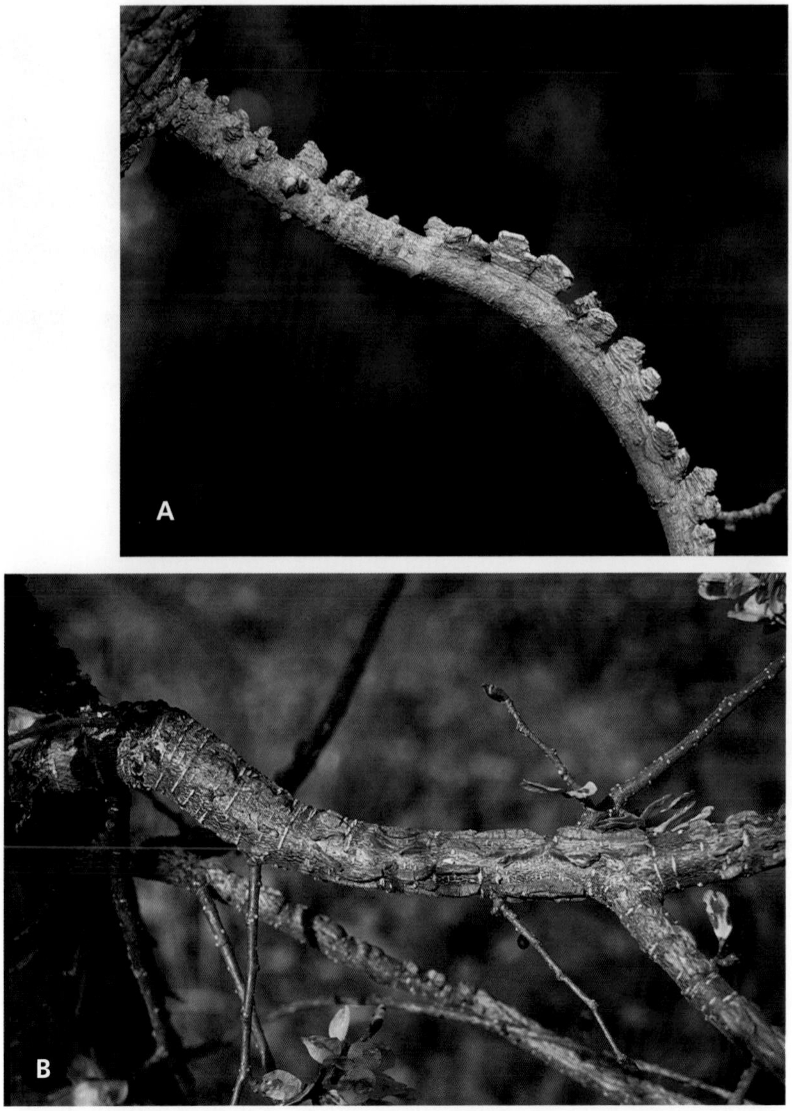

Figure 4.11. Corky bark is not uncommon on sweetgum (*Liquidambar styraciflua*) (A) and elm (*Ulmus* sp.) (B), and it is sometimes confused with insect damage or disease.

Galls, Burls, and Other Outgrowths

Various abnormal woody outgrowths occur on tree stems, branches, trunks and roots. These outgrowths are extremely variable in size and shape. They may be caused by fungi, bacteria, mycoplasmalike organisms, insects, mites, nematodes, and mistletoe, or they may be abiotic in origin.

The causes of abiotic outgrowths on tree trunks and branches are unknown.

Figure 4.12. Peeling bark on eucalyptus (*Eucalyptus* sp.) (A) and river birch *(Betula nigra)* (B). New bark beneath the peeling bark is smooth, healthy, and tightly attached to the wood.

Figure 4.13. Bark cracks often occur in spring and summer as a result of rapid trunk growth. They do not extend into the wood and close as the season progresses.

Although they may grow very large, sometimes completely surrounding a branch, these swellings apparently do no harm to the tree. Some are composed of a mass of adventitious or dormant buds that produce shoots for a short time. Many of the small shoots die, resulting in additional bud formation and growth until a dense mass of twigs is formed. As these outgrowths increase in size, the wood laid down each year at the bases of the aborted buds also becomes greater in size, resulting in a burl.

The distorted nature of the xylem cells in a burl makes the wood valuable for gun stock and furniture manufacture. Burls occur naturally on a number of species, including elm (*Ulmus* spp.), European white birch *(Betula pendula)*, honey locust *(Gleditsia triacanthos)* (fig. 4.14), maple (*Acer* spp.), oak (*Quercus* spp.), and coast redwood *(Sequoia sempervirens)*.

Galls or outgrowths covered by bark sometimes occur on trunks and branches (fig. 4.15). These swellings resemble burls, but they are basically different in their formation as they do not occur from bud proliferation. Instead, these smooth galls are caused by wood swelling beneath the bark. They vary in size, from small to very large, and sometimes encircle the entire trunk or branch. The causes of such galls are unknown.

Several species of eucalyptus (*Eucalyptus* spp.) develop basal woody outgrowths or swollen underground stems, or lignotubers (fig. 4.16). Lignotubers are normal growths formed by the plant as an adaptation to fire or other environmental stress. The growths contain latent buds that sprout after the treetop is killed or severely injured.

Figure 4.14. Burls, as found on this honey locust *(Gleditsia triacanthos)*, are considered to be harmless to the tree.

Figure 4.15. Gall on white mulberry *(Morus alba)* trunk. The cause of these galls is unknown.

Galls caused by certain bacteria, fungi, insects, mites, and nematodes may resemble abiotic galls. The bacterium *Agrobacterium tumefaciens* causes the disease crown gall, which appears as galls mainly on the root crown, at the soil line, or just below the soil surface. The galls may also grow on roots, branches, and the trunks of certain trees and shrubs. Young crown galls are smooth but become rough, brittle, and cracked as they age and increase in size. As opposed to abiotic galls, crown galls consist of soft, undifferentiated tissue that lacks bark and the annual rings of wood. A bacterial gall caused by *Pseudomonas syringae* affects oleander (*Nerium oleander*) and olive (*Olea europaea*). Galls may form on the stems of both species and on the leaves of oleander. Like crown galls, the bacterial galls have rough, warty surfaces. Western gall rust caused by the fungus *Peridermium harknessii* (or *Endocronartium harknessii*) causes round swellings or galls on the branches of two- and three-needle pine species in the western United States. In spring, the galls become covered with orange spores. Portions of the branch beyond the galls eventually weaken and die.

Certain soilborne bacteria (*Rhizobium* and *Frankia* spp.) can produce small growths, or nodules, on the roots of plants in the legume family and in several woody nonlegumes (*Alnus*, *Ceanothus*, *Casuarina*, and *Elaeagnus* spp.) (fig. 4.17). These beneficial bacteria enter the root hairs and cause the root cells to divide rapidly, forming the nodules. The bacteria live in the nodules in a symbiotic relationship with the plant, deriving food and minerals from the plant and in turn supplying the plant with nitrogen. Nitrogen-fixing nodules are usually found in bunches, mainly on younger roots, and may vary in size and shape. The cultivated annual legumes generally have large, sphere-shaped nodules, while nodules of biennial and perennial legumes are usually smaller, elongated, and in clusters. Nodules in nonlegumes are perennial.

Many plant galls are caused by certain species of insects and mites. These galls are the result of irritation caused by insect or mite feeding or when insects or mites inject plant growth regulating chemicals during egg-laying or feeding. Each insect or mite species produces its own distinctive gall. More galls occur on oak (*Quercus* spp.) than any other plant species. Galls caused by tiny wasps in the family Cynipidae are especially abundant on oaks.

Roots infected by root knot nematodes (*Meloidogyne* spp.) develop galls that resemble nitrogen-fixing nodules (fig. 4.18). Nematode galls vary in size from minute to very large. Heavily infected roots are often badly discolored and rotted due to secondary invasion by fungi and bacteria. Unlike legumes that have nitrogen-fixing root nodules, plants with nematode-infected roots become stunted and slowly decline in vigor.

Figure 4.16. Lignotubers at base of eucalyptus (*Eucalyptus* sp.) seedling. These swellings contain latent buds that sprout if the top of the tree is killed or severely injured.

Witches' Broom

Witches' broom is the proliferation of large numbers of shoots near the end of a branch in response to terminal bud death. The shoots in a witches' broom usually continue to grow, resulting in a dense cluster of branches. Brooms on conifers and angiosperms (seed-bearing plants such as oaks and roses) have been linked to mutations in

Figure 4.17. Nitrogen-fixing nodules on roots of berseem clover *(Trifolium alexandrinum)*, an annual legume (A). These nodules are formed by beneficial bacteria that enter the root hairs and cause root cells to divide rapidly. Red or pink tissue inside the nodule indicates that it is actively fixing nitrogen (B).

vegetative cells. Fungi, mycoplasmalike organisms, viruses, mistletoe, mites, insects, and herbicide phytotoxicity (especially from glyphosate) can also cause witches' broom. (see fig. 3.18, p. 27)

Fasciation

Fasciation is an abnormal flattening of stems that is caused by cell mutations in several woody species, including ash (*Fraxinus* spp.) and pittosporum (*Pittosporum tobira*) (fig. 4.19). Instead of a single growing point or apical bud on a stem, a jagged row of growing points develops, resulting in a wide, flattened stem. The flattened stem has a ribbed appearance, as though several stems are fused. Because the stems grow at different rates, fasciated shoots are usually coiled and twisted. Densely packed buds along each stem in the group may produce undersized leaves. Fasciated growth may revert to normal growth. The causes of the mutations that result in fasciation are unknown.

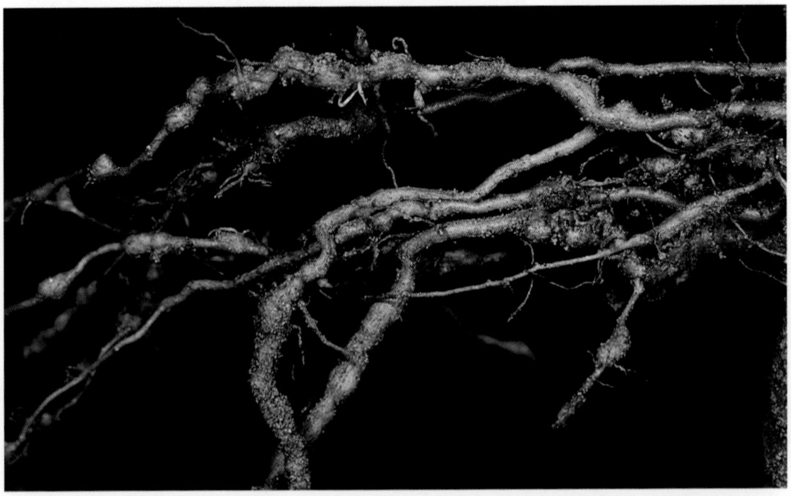

Figure 4.18. Root knot nematode galls can be confused with nitrogen-fixing nodules. Nematode-infested roots on this annual plant are swollen, stunted, and distorted, causing a decline in vigor.

Lichens

Lichens are small perennial plantlike organisms that consist of a fungus and a green or blue-green alga growing in a symbiotic relationship. Lichens colonize various surfaces, including rocks, fence posts, and often tree bark. Many species grow rapidly in the presence of sunlight and seasonal moisture, covering the trunks and branches of declining or dead trees (fig. 4.20). While they may be associated with declining trees, they are not

Figure 4.19. Fasciation in pittosporum *(Pittosporum tobira)* appears to be a serious disorder. This abnormal flattening of the stem occurs as a result of cell mutations (cause unknown) in several woody species.

Figure 4.20. Lichen on oak (*Quercus* sp.) (A) and moss on California bay laurel *(Umbellularia californica)* (B) are epiphytes that are generally harmless to the tree.

parasitic. Lichens use the bark only as a surface on which to grow. They are not as likely to develop on rapidly growing trees, because new bark is being formed faster than the lichens can grow.

Epiphytes

Epiphytes are photosynthetic plants that grow on tree trunks and branches. Examples are mosses, which are common on trees in cool, northern areas, and members of the bromeliad and orchid families in warm and tropical regions. Epiphytes grow in both full sun and shade. Spanish moss *(Tillandsia usneoides)* is an example of a common moss epiphyte that grows rapidly in full sun, especially on declining trees (fig. 4.21). Epiphytes are not parasitic, but use the tree only as a growing surface.

Figure 4.21. Spanish moss *(Tillandsia usneoides)* growing on oak (*Quercus* sp.) is an epiphyte that is often more plentiful on trees of low vigor.

CHAPTER 5

Specific Abiotic Disorders

Symptoms, Causes, Look-Alike Disorders, Diagnosis, Sensitive and Tolerant Species, and Remedies

WATER DEFICIT (WATER STRESS, DROUGHT, DEHYDRATION)

Almost every aspect of plant growth and development can be affected by a water deficit, including a plant's anatomy, morphology, physiology, and biochemistry. Deficits occur when water loss exceeds supply, that is, when transpiration is greater than water uptake.

Symptoms

Water deficit symptoms range from slow growth to death of the whole plant. The level of injury depends on the severity and duration of the deficit and the sensitivity or tolerance of the plant. Symptoms can be grouped into two categories: those associated with acute water deficits, and those associated with chronic deficits. Acute deficits are short-lived (hours to days), while chronic deficits last from a few days to several months. Both can cause injury symptoms that range from mild to severe.

Acute Deficit

When water loss increases substantially over a short period, or when water supply declines rapidly, plant tissues become dehy-

Figure 5.1. Wilt is an early symptom of water deficit. Leaves and shoots of periwinkle *(Vinca major)* have lost turgidity due to warm air temperatures and dry soil conditions (A). Wilting leaves of European hackberry *(Celtis australis)* during a hot summer day are indicative of a water deficit (B).

drated. Initially, leaves and shoots of many species wilt (fig. 5.1). Some species may wilt during the warmest part of the day but rehydrate in the evening (incipient wilt) (fig. 5.2). As fiber content increases during the growing season, leaves and shoots are less prone to wilt.

Figure 5.2. On warm summer days, leaves of poor man's rhododendron *(Impatiens oliveri)* wilt even when soil moisture is plentiful. This is called *incipient wilt.*

Figure 5.3. Marginal necrosis is a common symptom of water deficit, as seen on these sawleaf zelkova *(Zelkova serrata)* leaves.

If a water deficit continues, tissues may dehydrate to the point of becoming necrotic. Leaf necrosis may be expressed as a marginal burn, tip burn, or as irregular areas of dehydration in the leaf blade (fig. 5.3). Often, rapid dehydration causes the leaves to turn reddish brown, with distinct borders between hydrated and dehydrated tissue (fig. 5.4). In some species, extensive leaf drop may occur (fig. 5.5). If an acute water deficit is severe, the whole plant may die.

Chronic Deficit

Chronic water deficit causes slow growth or stops growth altogether. This symptom is frequently difficult to identify, however, since long-term observation of the species and the individual plant is needed. Comparing plants of similar age in similar growing conditions is usually required to assess a growth rate decline (fig. 5.6).

In addition to reduced growth rate, leaf size is often reduced on water-stressed plants (fig. 5.7). Leaf color may be less intense (e.g., change from deep green to

Figure 5.4. Acute water deficit may cause red-brown necrosis (with distinct margins) on the leaves of many species.

gray-green or blue-green), and certain species drop leaves prematurely, such as tulip tree *(Liriodendron tulipifera)* and southern magnolia *(Magnolia grandiflora)*. The foliage of deciduous species may show fall colors much earlier than normal (fig. 5.8).

Moderately severe prolonged water stress causes shoot and branch dieback (fig. 5.9) and, in some cases, bark cracking and trunk bleeding. Chronically water-stressed plants often have low pest resistance, and secondary injury from insects or pathogens is common in some species. For

Figure 5.5. If a water deficit is severe, leaf drop and branch dieback will be evident in sensitive species.

Figure 5.6. Chronic water deficit may be expressed as a reduction in growth. These chrysanthemum plants *(Chrysanthemum morifolium)* were watered (left to right) every 10 days, 4 days, every day, and twice a day. Plants receiving lesser amounts of water produced less growth.

Figure 5.7. Knowing typical leaf size and color helps diagnose water deficit injury. Water-stressed foliage of coast redwood *(Sequoia sempervirens)* exhibits small needles and a dull green color (two shoots on left) when compared to foliage of a well-irrigated plant (right).

Figure 5.8. Water deficits may lead to early fall color in deciduous species. This Chinese pistache *(Pistache chinensis)* is showing fall color in August. The tree is located in a nonirrigated site with high evaporative potential.

example, bark beetle injury in Monterey pine *(Pinus radiata)* is common when trees in warm summer areas are inadequately irrigated (fig. 5.10).

Figure 5.9. Prolonged water deficit causes branch dieback in certain species. Canopy thinning and branch dieback occurred in this southern magnolia *(Magnolia grandiflora)* when turf irrigation was discontinued.

Figure 5.10. When water-stressed, Monterey pine *(Pinus radiata)* is highly susceptible to injury from bark beetles. The trees shown on left were not irrigated in the summer months and were killed by California five-spined engraver beetle *(Ips paraconfusus)*.

Causes

Often a combination of conditions or multiple factors contribute to water stress. Plant water deficits can be caused by conditions that

- limit the supply and availability of water in the soil (soil and water factors)
- affect water loss from plants (aboveground factors)
- affect a plant's capacity to absorb or retain water (plant factors)

Soil and Water Factors

Soil moisture levels are rarely optimal. Even in irrigated landscapes, inadequate water supply and availability frequently contribute to water deficits (fig. 5.11). Water supply to the root zone can be limited in a number of ways:

- Below-normal rainfall. California's weather pattern of wet winters and dry summers creates a prolonged period of low soil moisture. In years when rainfall is below normal, nonirrigated plants may

Figure 5.11. Water deficits are often caused by inadequate irrigation. These European white birch *(Betula pendula)* experienced severe water deficit and dieback when turf irrigation was discontinued. Note that junipers (*Juniperus* spp.) growing next to the birch are not showing similar symptoms.

deplete the supply of available water before it is replenished in the fall.

- Irrigation system design and function. Typical problems with irrigation systems include poor design (inadequate coverage, low water pressure) and poor function (broken valves, timer, heads, etc.).
- Irrigation scheduling. Typical problems include insufficient water volume or frequency and an application rate that is much greater than the infiltration rate, causing runoff.
- Low infiltration rate. Even when adequate quantities of water are applied, sufficient water may not infiltrate the plant root zone. The infiltration rate can be inherently low (e.g., in fine-textured soils), or it can be reduced by compaction, poor soil structure, slopes, surface barriers (hardscape or heavy organic mulch layers), and fill soils.

Water availability in the root zone can be limited when

- a small volume of water (per unit of soil volume) is retained in the soil. This often occurs in soils with low water-holding capacity (sand, loamy sand, and container soils) and in hydrophobic soils.
- the volume of soil in the root zone is small compared to the plant canopy. Soils that often have small root zone volumes include container soils, shallow soils, compacted soils, and soils limited by infrastructure elements (e.g., retaining walls, foundations, roadways, utility boxes, etc.).

To determine whether supply or availability are contributing to plant water deficits, assess the soil physical properties (texture, structure, depth, and bulk density) and, in irrigated landscapes, evaluate the irrigation system design and operation.

Aboveground Conditions

Environmental factors such as light intensity, wind, temperature, relative humidity, and aspect determine the evaporative potential of a site. Plants in sites with high evaporative potential are more likely to suffer water deficits than those in sites with evaporative potential (fig. 5.12). For example, an isolated tree exposed to full sun in a windy parking lot is more likely to experience water deficits than a tree of the same species planted close to other trees and protected from wind.

Plant Factors

Although any plant can be injured by a water deficit, certain species are more sensitive than others. Plants from temperate zones (e.g., maple and horsechestnut) have a higher potential for injury than species from xeric (dry) zones such as century plant (*Agave* spp.) and mesquite (*Prosopis* spp.). Adaptations to control water loss or enhance absorption potential are found in species that are regularly exposed to water deficits. These adaptations include extensive root systems, small leathery leaves, and gray-green leaf color. Species without such adaptations are prone to injury, such as ferns, azalea (*Rhododendron* spp.), and hydrangea (*Hydrangea macrophylla*) (see

Figure 5.12. Trees planted in sites with high evaporative potential and limited soil volume are particularly prone to water deficit. Reflected light and heat from cars and pavement, restricted soil volume, and no irrigation have contributed to severe water deficit in this ash (*Fraxinus* sp.).

"Sensitive and Tolerant Species," p. 58).

Plants with limited root development, such as young plants and newly planted container stock, are also prone to water stress (fig. 5.13). Ball-and-burlap stock and bareroot stock may also have root development limitations that increase their sensitivity to water stress. Trees or large shrubs that have lost roots during transplanting are also prone to stress (fig. 5.14). In addition, plants growing in suboptimal conditions (e.g., flooded sites, compacted soils) are likely to have a reduced capacity for water absorption.

Plants that have large leaf areas relative to the soil volume their roots occupy are likely to experience water deficits. For example, large plants in small containers have a much higher potential to suffer water stress than small plants in large containers (fig. 5.15).

Plants that have sustained root injury have a high potential for water stress. Root cutting that occurs during construction or utility line installation frequently leads to water stress symptoms (fig. 5.16).

Figure 5.13. Newly planted container stock is prone to water deficit. Limited amounts of water held in the root ball can be lost rapidly under high evaporative conditions. These trees were not irrigated as frequently as needed to avoid water deficit.

Figure 5.14. Plants that have had roots cut (intentionally or unintentionally) are subject to water deficit. This coast live oak *(Quercus agrifolia)* was transplanted from another site. In the process, much of the root system was cut, and water deficit followed. Symptoms include extensive leaf drop and branch dieback.

Figure 5.15. Plants growing in containers are commonly subject to water deficit due to restricted soil volume and low water-holding capacity. Water deficit symptoms in this relatively large Indian laurel fig *(Ficus microcarpa)* are typical of many plants in containers.

Look-Alike Disorders

Although injury from water deficit can be distinctive, it can also be confused with a number of other disorders. This is particularly the case with chronic water deficit. Look-alike disorders include

- salt injury
- aeration deficit
- specific ion toxicity (particularly chloride and boron)
- gas injury
- sunburn
- wind
- herbicide injury (particularly contact herbicides and nonselective materials)
- root disease and vascular wilt pathogens (e.g., *Phytophthora* spp., *Verticillium dahliae*, *Fusarium* spp., *Ophiostoma ulmi*, and *Erwinia amylovora*)
- insect injury (bark beetles, borers, root weevils)
- wildlife injury (gophers, voles, squirrels, deer, mice)
- mechanical injury (to root or trunk tissues)

Figure 5.16. Construction activities can have severe impacts on established trees. Trenching next to this tree removed many roots, and water deficit injury is likely to follow.

Species that naturally shed leaves earlier than other species, such as honey locust (*Gleditsia triacanthos*) and California buckeye (*Aesculus californica*) may be mistakenly diagnosed as being injured by water stress. In addition, deciduous species that retain leaves into the winter (e.g., pin oak, *Quercus palustris*) may appear to be water-stressed (see chapter 4).

Diagnosis

Visual assessment and specialized diagnostic instruments can be used to assess water deficits.

Visual Assessment

Plant factors

- Identify the plant. Is the species sensitive to water deficit?
- Consider all symptoms and determine whether they are consistent with acute or chronic water stress.
- Check the condition of nearby plants. Do they show similar injury symptoms? Are they the same species?
- Evaluate plant growth (measure the annual growth increment). Considering site conditions and plant age, is growth normal for the species?
- Consider the plant history. Have there been impacts that may affect plant water uptake (e.g., root cutting, soil compaction)? Look for new features such as lights, sidewalks, utility boxes, swimming pools, etc.

Site factors

- Evaluate the soil. Is it dry, compacted, or shallow? Are there barriers to water infiltration on the surface (concrete, asphalt, fill soil, etc.)? Are the plants on a slope?
- Check the irrigation system. Is it operating properly? Is the irrigation schedule adequate? Is the coverage sufficient? Is water getting into the root zone?
- Consider the evaporative potential of the site. Is it windy, hot, or dry (low humidity)? Is there reflected light? Is the exposure south, west, or southwest?

Diagnostic Instruments

Plant water status

Leaf or stem water potential can be measured using a pressure chamber (fig. 5.17). The chamber measures tension levels in xylem sap. High tension measurements indicate water stress. Tension levels vary among species, however, and baseline data on normal and critical tensions (indicating water stress) are needed to interpret readings. Although used mainly in the past for research, this instrument has been adapted for field use.

Figure 5.17. Water deficits can be assessed by measuring leaf or stem water potential using a pressure chamber. Previously used for research, this instrument has been modified for use in the field.

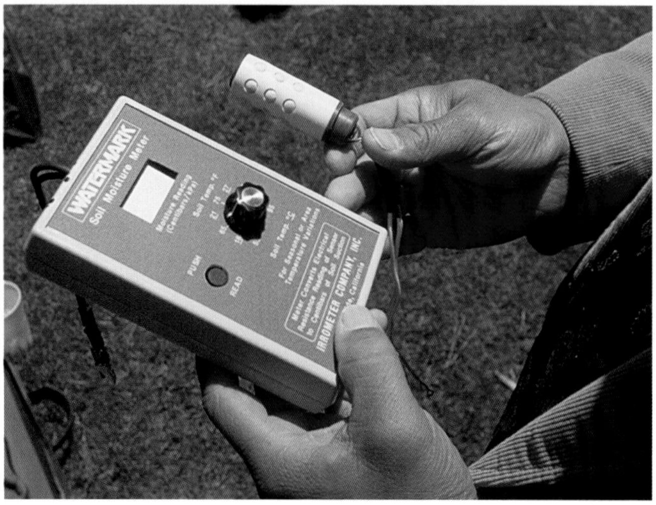

Figure 5.18. Soil moisture level can be assessed by measuring matric potential with electrical resistance sensors. The sensor is buried in the soil, and resistance is read using a hand-held digital meter.

Soil water status

Soil moisture status can be assessed by measuring soil matric potential using gypsum blocks and tensiometers (fig. 5.18). To properly interpret readings, it is necessary to understand the moisture release characteristics for the soil. For example, sands and container soils can be dry at relatively low tensiometer readings, while equivalent readings in clay indicate a well-hydrated condition.

Sensitive and Tolerant Species

Species selection is an important way to prevent water deficit injury (table 5.1; fig. 5.19). Many plant manuals provide information on the relative water needs of landscape species. An extensive list of species and irrigation-need evaluations is presented in *A Guide to Estimating the Irrigation Water Needs of Landscape Plantings in California* (Costello and Jones 2000).

Remedies

Before developing a plan for remediation, accurately identify the cause of the water deficit. Remedies should take into consideration whether the cause is related to the soil, plant, irrigation, or site.

Soil

- Assess soil physical properties to determine whether they are limiting factors. Improve the soil condition that limits water availability or supply. Cultivate to reduce compaction, amend to increase water-holding capacity, treat hydrophobic soils with a wetting agent, increase soil volume where limited, or improve infiltration rate (e.g., by using a hollow-tine aerator).
- Add mulch to the soil surface to improve soil structure, decrease evaporation, and reduce competition from weeds.

Irrigation

- Correct irrigation system malfunctions (timer, valves, heads, etc.).
- Improve system uniformity. Conduct a "can test" to evaluate water distribution over the irrigated area (see Pittenger 2002, p. 73).

TABLE 5.1. Water needs of selected landscape plants

Scientific name	Common name
High water needs	
Acer palmatum	Japanese maple
Alnus spp.	alder
Betula spp.	birch
Fuchsia spp.	fuchsia
Hydrangea macrophylla	hydrangea
Impatiens spp.	impatiens
various genera	ferns
Low water needs	
Acacia spp.	acacia
Arbutus menziesii	madrone
Arctostaphylos spp.	manzanita
Nerium oleander	oleander
Olea europaea	olive
Pinus canariensis	Canary Island pine
Pinus halepensis	Aleppo pine
Pinus sabiniana	digger pine
Prosopis spp.	mesquite
Quercus agrifolia	coast live oak
Quercus douglasii	blue oak
Quercus wislizenii	interior live oak
various genera	iceplant
various genera	palms

Figure 5.19. Species with dissimilar water needs should not be planted in the same irrigation zone. Here, a tulip tree *(Liriodendron tulipifera)* planted in manzanita (*Arctostaphylos* sp.) groundcover became severely water-stressed.

- Adjust the irrigation schedule to supply the amount of water the plants need. Develop a water budget.
- Provide sufficient water to thoroughly wet the soil in the root zone (deep irrigation, drip, hand-watering, etc.).
- Match the water application rate with the infiltration rate of soil.
- Adapt the irrigation system and schedule for plants on slopes. For example, use short run times and repeat cycles.

Plant

- Select species that are appropriate for the site and environmental conditions. Group species with similar water needs in an irrigation zone (hydrozone).
- Irrigate at a level and frequency appropriate for the plants in a hydrozone.
- Improve root zone conditions that have impaired the function of the root system (e.g., improve drainage in saturated or flooded soils).
- Avoid injuring root systems during construction activities. Do not trench across root systems; use excavation techniques that avoid root cutting (e.g., hydro-excavation or pneumatic excavation).
- Provide sufficient root space for plants. Do not overplant an area with limited soil volume.

Site

- Evaluate the site to determine what is increasing the evaporative potential. Modify the site environment to reduce the evaporative potential (e.g., add shade, increase humidity, protect from wind, reduce light reflection, reduce heat inputs).
- Select species that are well adapted to the site.
- Adjust the irrigation to match the evaporative potential of the site.

Aeration Deficit (poor aeration, anoxia, oxygen deprivation, oxygen stress, wet feet)

Plant roots require an adequate supply of oxygen for growth and development. Insufficient supply or availability of oxygen causes aeration deficits and limits root function. Root respiration is usually the first plant process to be restricted, followed by disruptions in metabolism, nutrient uptake, water absorption, and photosynthesis. Aeration deficits commonly occur in irrigated landscapes.

Symptoms

Aeration deficits can produce a range of symptoms from slow growth to death of the whole plant. The severity of the symptoms depends on species tolerance or sensitivity, severity and duration of the aeration deficit, and the initial health or condition of the plant. As with water deficits, aeration deficits can be acute or chronic.

Figure 5.20. Acute aeration deficit can cause whole-plant death. Excess irrigation and poor drainage excluded oxygen in the root zone of this newly planted photinia *(Photinia fraseri).* Symptoms include leaf necrosis, branch dieback, root discoloration, blackening of soil, and foul odor.

Acute Deficit

When oxygen supply or availability is below a critical level for a short time (hours to days), an acute deficit occurs. Symptoms include wilting, extensive leaf drop, and, in some cases, death of the whole plant (fig. 5.20). Roots suffocate and no longer absorb nutrients or water. Frequently, they appear discolored and water-soaked (fig. 5.21; see also fig. 3.26, p. 29). In some species, wood tissues in the stem or trunk may become discolored.

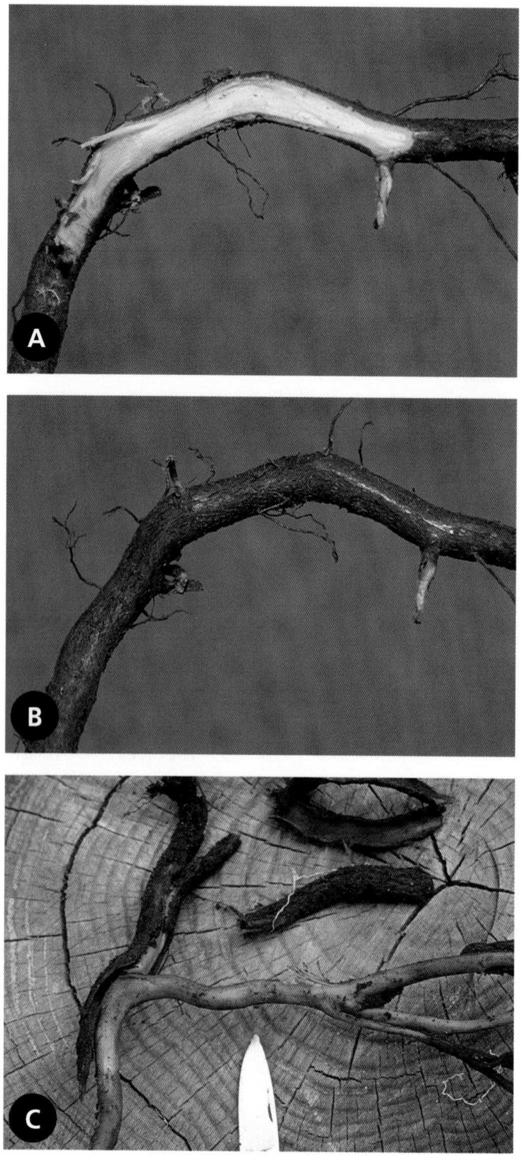

Figure 5.21. Healthy roots appear white and turgid (A), and the bark is firm and intact (B). Roots injured by an aeration deficit can appear discolored and water-soaked, and bark can lift off easily (C).

Figure 5.22. Slow growth, canopy thinning, and chlorosis in this Victorian box *(Pittosporum undulatum)* were caused by an aeration deficit. The aeration deficit resulted from excess irrigation and poor drainage. Previously, several India hawthorn *(Rhaphiolepis indica)* planted at the base of the Victorian box were killed by the aeration deficit.

Plants exposed to acute aeration deficits are more susceptible to root disease. Frequently, pathogens such as *Phytophthora cinnamomi* (water mold root rot) or *Armillaria mellea* (oak root fungus) infect roots and cause further injury.

Chronic Deficit

An aeration deficit that persists over an extended period (weeks to months) is considered to be chronic. Chronic, mild deficits initially slow plant growth; shoot length and leaf size are smaller than normal. This condition may be difficult to identify, however, since knowledge of the normal growth rate for the species is needed.

As a chronic deficit continues, leaves become chlorotic, and cankers or bleeding may occur. Typically, older leaves are affected first, then younger leaves. Plants begin to look anemic, or "sick." Leaf drop and canopy thinning is common (figs. 5.22–23). Lenticels may develop on the stem, and adventitious roots may appear near the root crown. In some species, cankers may be seen on the stem or trunk, and resinous or gummy substances may exude from the cankers (fig. 5.24A). Bark may split, exposing discolored wood beneath.

In more severe cases, twigs and small shoots die, followed by death of an entire branch. Eventually, a section of the canopy is affected or the whole plant may die (fig. 5.24B–C).

Under anaerobic (no oxygen) conditions, soil and roots smell like rotten eggs because gas containing sulfur is produced. When the condition persists, the soil color may change to bluish gray or black (fig. 5.25).

As with acute deficits, plants suffering from chronic aeration deficits are more sus-

Figure 5.23. Conifers injured by aeration deficits exhibit symptoms similar to broadleaf plants: slow growth, needle drop, and chlorosis. Accumulation of water at the bottom of this slope led to the decline of Monterey pine *(Pinus radiata)*. Note that Monterey pines on the slope are not showing symptoms. All trees were the same size and age at planting.

Figure 5.24. In some species, trunk bleeding and cankers can result from aeration deficits (A). When aeration deficits are prolonged and severe, extensive canopy dieback can be found. Localized poor drainage caused an aeration deficit in the root zone of the southern magnolia *(Magnolia grandiflora)* in the foreground (B). After 2 years, the same tree was virtually dead (C). Drainage in the root zone of the healthy tree in the background was substantially better than that of the injured tree.

Figure 5.25. Blackened soil is an indication of anaerobic soil conditions.

ceptible to root disease and other plant pathogens. Infections can cause further injury, masking symptoms of the primary cause (aeration deficit).

Causes

Plant aeration deficits are caused either by critically low oxygen concentrations in the root zone or by substantial reductions in oxygen movement through the soil (oxygen diffusion rate). Oxygen concentration and diffusion rate are in turn affected by a variety of soil and plant factors.

Low Oxygen Concentration

Oxygen depletion occurs when oxygen consumption in the root zone exceeds supply. Oxygen is consumed during respiration. Roots respire as they metabolize and grow, and soil microorganisms respire as they decompose organic matter. The combined respiration of roots and microorganisms reduces oxygen concentration. Consequently, soils that have active root systems or are high in organic matter generally have higher respiration levels than those with little root activity or low organic matter.

When respiration is high and oxygen

supply becomes limited, concentrations of oxygen in the root zone can be reduced to injurious levels. Oxygen concentration in the atmosphere is approximately 21 percent. Although most plants are not injured until the oxygen concentration in the root zone declines to below 10 percent, sensitive plants may be injured at between 10 and 15 percent. As the oxygen concentration declines below 10 percent, root functions are increasingly harmed. Notably, however, certain species (e.g., rice and other flood-tolerant species) are adapted to tolerate very low levels of oxygen (less than 5 percent).

Soil temperature has an important effect on injuries caused by low oxygen concentration. Plants are generally less tolerant of low oxygen levels when the soil temperature is relatively high; they are more tolerant at lower temperatures. This is because respiration increases as temperature increases and declines as temperatures drop. Increased respiration during warmer soil temperatures reduces the oxygen concentration, making oxygen deficits more likely.

The supply of oxygen from the atmosphere to the root zone can be limited by physical barriers at or on the soil surface, such as asphalt, concrete, plastic, and surface compaction (fig. 5.26). These barriers can reduce the rate of oxygen entry into the soil.

Low Oxygen Diffusion Rate (ODR)

Oxygen moves through soils by diffusion, that is, from areas of high concentration to areas of low concentration. As oxygen is consumed by roots or microorganisms, its concentration is reduced. This establishes a concentration gradient for oxygen diffusion from the atmosphere through the soil to the respiring surface (root or microorganism). The greater the oxygen gradient, the higher the diffusion rate.

A soil oxygen diffusion rate (ODR) greater than 0.3 $\mu g/cm^2/min$ is sufficient to meet the respiratory needs of most species. As ODR declines below this level, however,

Figure 5.26. Surface materials such as concrete or asphalt are thought to cause aeration deficits by reducing the rate of oxygen movement into soils. Pavement around this blackwood acacia *(Acacia melanoxylon)* may have reduced aeration levels and contributed to its decline (A). There are many cases, however, where trees do not appear to be injured by pavements, such as these pines (*Pinus* spp.) in Tahoe City, California (B).

Figure 5.27. Soil compaction caused by foot traffic resulted in a reduction in air-filled pore space around these trees. This limits the oxygen diffusion rate in soils and can cause aeration deficits.

oxygen supply to root surfaces is diminished, and aeration deficits may occur: not enough oxygen is arriving at root surfaces to meet respiratory needs.

The oxygen diffusion rate is reduced by barriers along the diffusion pathway (the "air-filled pore space" in soil). This pathway is a network of pores that are large enough to be occupied by air after water drainage (macropores) that serve as channels for gas diffusion. A soil with an air-filled pore space of 10 to 12 percent (by volume) is satisfactory to meet the ODR needs of most plant species. Soils with good structure or soils with inherently large pore space (e.g., sand, loamy sand, and container soils) generally have 10 to 12 percent (or more) air-filled pore space. Soils with little air-filled pore space (e.g., fine-textured soils such as clay or poorly structured soils) are prone to low ODR levels. In addition, soils that have experienced a reduction in macropore space (e.g., by compaction) have diminished ODR (fig. 5.27). When ODR is diminished below a critical level for a species, then an aeration deficit occurs.

Water is perhaps the most common barrier to oxygen diffusion. Oxygen diffuses through water 10,000 times slower than through air. When water occupies macropore space, oxygen diffusion is reduced substantially. This occurs in flooded soils, poorly drained soils, and soils that are irrigated excessively. It is particularly a problem in soils that typically have little macropore space (e.g., clay, compacted soils, or poorly structured soils). In these soils, water occupies the space needed for gas diffusion, and oxygen cannot move to root surfaces fast enough to meet respiratory needs (fig. 5.28). As a result, an aeration deficit occurs.

Because oxygen moves from the atmosphere into the soil to the roots, barriers on the soil surface such as asphalt, concrete, fill soil, and surface compaction may reduce oxygen movement into the soil (fig. 5.29). If so, ODR decreases and aeration deficits may follow. It is not known by how much ODR is reduced by these barriers, however. Although many cases have been observed where it is believed that plants have been injured by surface barriers, there are many more cases of plants growing very well in spite of being surrounded by surface barriers (e.g., trees planted in sidewalks). This may be because most paved surfaces have many fine cracks through which gases may diffuse.

Look-Alike Disorders

Aeration deficit injury is not highly distinctive and can be confused with other disorders, such as chronic water stress, salt injury, gas injury, high light injury, nitrogen deficiency, air pollution, herbicide injury (soil-residual herbicides that cause chlorosis in older foliage), and root diseases such as water mold root rot (*Phytophthora* spp.) or oak root fungus (*Armillaria* spp.).

Diagnosis

Diagnosis of aeration deficit injury requires familiarity with symptoms of acute and chronic deficits and an understanding of soil conditions that may limit oxygen concentration or movement in the soil. Close visual inspection and, in some cases, measurements of soil oxygen status will be needed for accurate diagnosis. Review the symptoms of aeration deficit (above) and determine whether the plant's symptoms are consistent with them. Also:

- Check the soil moisture content. This can be done using a shovel, soil sampling tube, steel probe, or tensiometers. Is the soil very wet? Have there been large water inputs from rainfall? Is there an irrigation system? If so, what is the irrigation schedule? Is it too long or too frequent? Is there a drying period between irrigations? Are there irrigation system leaks?
- Evaluate the color and smell of the soil. Does it smell like rotten eggs? Is it a bluish gray color?
- If possible, sample the roots. Do they appear water-soaked or discolored? Do they smell like rotten eggs?
- Evaluate surface drainage patterns at the site. Is surface water being directed to the affected plants? Does water form puddles (ponding) in the area? Is water being channeled to the affected plants (e.g., from downspouts connected to a roof drainage system)?

Edema (Corky Scab, Corky Excrescence, Scab, or Scurf)

Edema (or oedema) is the growth of small, raised, corky outgrowths resembling lenticels on the underside of leaves. It is often seen in camellia *(Camellia japonica* and *C. sasanqua),* hibiscus (*Hibiscus* spp.), ivy (*Hedera* spp.), privet (*Ligustrum* spp.) and yew (*Taxus* spp.). Edema may be caused by aeration deficits; it typically develops in plants growing in waterlogged soils, especially when transpiration is reduced, such as during cloudy weather. Edema also develops under conditions of exceptionally high humidity, such as in greenhouses. The wartlike growths usually have a corky texture and are light brown or rust-colored (fig. 5.30). They may occur in small groups or cover large areas of a leaf. The upper surfaces of leaves with edema are normal in size and texture but are often chlorotic. The corky eruptions may be mistaken for rust diseases caused by fungi.

Figure 5.28. Air-filled pore space becomes water-filled pore space in flooded soils. Water physically excludes air, greatly reducing oxygen diffusion. Rainfall combined with poor drainage causes these planters to fill with water leading to an aeration deficit and decline of cork oak *(Quercus suber)* (A). Water delivered from the downspout likely contributed to an aeration deficit and subsequent decline of this evergreen pittosporum *(Pittosporum crassifolium)* (B).

Figure 5.29. Fill soil may contribute to aeration deficits and plant injury. The decline of this Aleppo pine *(Pinus halepensis)* was attributed to an aeration deficit caused by the placement of approximately 2 feet (0.6 m) of fill around its base (A). The fill soil covered the trunk flare (B).

- Evaluate the soil's internal drainage. Are there barriers to internal drainage such as hardpan? Is there a perched water table near the surface? Are there tidal water effects at the location? Are there subterranean barriers such as utility boxes, basements, transit lines, or sewer lines? Is the soil fine-textured or poorly structured (i.e., does it have a low capacity for internal drainage)? Is the sodium content of the soil high?
- Check the bulk density of the soil. If compaction has occurred, bulk density will be above a critical level for the soil texture class. Use a field-core-sampling tool or use the volume excavation technique (see Lichter and Costello 1994). Is bulk density at or above a critical level (for values, see Harris, Clark, and Matheny 1999)?
- Determine whether the surface grade of the soil has been changed. Look for indications that fill soil has been added on top of field soil around affected plants. For trees, look for a root flare (buttress roots) at the base of the trunk. If a root flare is not present (i.e., if the trunk arises from the ground like a pole), it is likely that fill soil has been added (fig. 5.31).
- Look for surface barriers. Is concrete, asphalt, or other hardscape present near the affected plants? Are the barriers sealed such that water or air may not be able to enter the soil? Have they recently been installed? How much of the root zone has been covered?
- Evaluate soil organic matter content. Is it very high (greater than 20 percent)? Is the organic matter composed of slowly decomposing materials (e.g., wood chips) or quickly decomposing materials (e.g., leaves)?
- Evaluate the oxygen diffusion rate and/or the oxygen concentration. Although it is possible to measure the oxygen status of the soil, there are practical limitations. Measuring ODR requires equipment that has not been adapted for diagnostic use. The equipment is useful for research, but general field use would require substan-

tial time, skill, and expense. Equipment to measure oxygen concentration is available, but variation in quality does exist. Follow established protocols for taking oxygen measurements. Remember that an ODR deficit may occur even when the oxygen concentration is satisfactory.

Sensitive and Tolerant Species

Although lists of aeration-deficit sensitive and tolerant species are not available, lists of flooding-sensitive species have been developed (table 5.2). It is possible that species sensitive to flooding are also sensitive to other types of aeration deficits (e.g., reduction in oxygen diffusion rate caused by changes in grade), but this has not been established. Consequently, the species that are sensitive to or tolerant of flooding listed here should not necessarily be considered as sensitive or tolerant to other types of aeration deficits.

Keep in mind that flood tolerance depends on the conditions to which plants have adapted. For example, after growing in drained soil, rice can be severely injured if flooded. Also, species tolerance varies with the time of the year. Dormant plants are more tolerant of low aeration conditions than actively growing plants. For example, valley oak *(Quercus lobata)* can tolerate flooding during the winter, but not during the summer.

Remedies

Identify the cause of an aeration deficit before developing a plan for remediation. Remedies are usually specific to the cause; for example, if the cause is a reduction in oxygen concentration, the remedy should focus on increasing the oxygen concentration.

Oxygen Concentration

Low oxygen concentration can be linked to organic matter content in the soil, surface barriers, and soil temperatures.

Figure 5.30. Small, raised, corky outgrowths (edema) on leaves of certain species such as this *Eucalyptus* sp. can result from a flood-induced aeration deficit.

Figure 5.31. If the root crown or trunk flare is not visible, fill soil is likely to have been placed around the base of the tree.

- If organic matter content is high (increasing respiration and oxygen use), reduce inputs of organic matter.
- If surface barriers are present, consider methods to improve the entry of air into the soil, including vents, barrier-free areas, or creating openings in the barrier.
- If high soil temperatures contribute to aeration deficits, add an organic mulch to the soil surface to moderate temperatures.

TABLE 5.2. Flooding tolerance of selected landscape plants

Scientific name	Common name
Flood-sensitive species	
Acer platanoides	Norway maple
Cercis canadensis	eastern redbud
Cornus florida	flowering dogwood
Crataegus × lavallei	Carriere hawthorn
Crataegus phaenopyrum	Washington thorn
Magnolia soulangiana	saucer magnolia
Malus spp.	flowering crabapple
Picea abies	Norway spruce
Picea pungens	blue spruce
Prunus persica	peach
Prunus serotina	black cherry
Prunus subhirtella	flowering cherry
Quercus agrifolia	coast live oak
Quercus kelloggii	black oak
Quercus lobata	valley oak
Robinia psuedoacacia	black locust
Thuja occidentalis	American arborvitae
Flood-tolerant species	
Acer negundo	box elder
Acer rubrum	red maple
Alnus spp.	alder
Betula spp.	birch
Celtis occidentalis	common hackberry
Fraxinus americana	white ash
Fraxinus pennsylvanica var. *lanceolata*	green ash
Gleditsia triacanthos var. *inermis*	thornless honey locust
Juglans nigra	black walnut
Liquidambar styraciflua	American sweetgum
Metasequoia glyptostroboides	dawn redwood
Morus alba	white mulberry
Nyssa sylvatica	tupelo
Platanus occidentalis	American sycamore
Populus spp.	poplar
Quercus bicolor	swamp white oak
Quercus palustris	pin oak
Salix spp.	willow
Sequoia sempervirens	coast redwood
Taxodium distichum	bald cypress
Tilia cordata	little-leaf linden
Ulmus americana	American elm

Oxygen Diffusion Rate

The oxygen diffusion rate can be improved by increasing the air-filled porosity of the soil, reducing moisture content, or correcting surface barrier problems (see Harris, Clark, and Matheny 2003, table 7.4).

Improving air-filled porosity

Cultivation, amendments, and mulches can improve air-filled porosity (fig. 5.32).

- Cultivation of compacted soils creates macropore space. This effect may be short-lived in some cases, however, because soil particles may return to precultivation compaction levels (e.g., following irrigation or rainfall). Eliminating compaction sources (traffic, equipment, etc.) and adding organic matter helps to retain cultivation effects.
- Organic amendments such as wood chips, sawdust, and bark improve air-filled porosity. These materials retain small amounts of water relative to other organic amendments such as peat. Generally, it is recommended that these organic materials be incorporated in the root zone at a rate of 25 to 50 percent of the soil by volume. This may not be practical over large areas, however, and incorporation may cause substantial root damage to existing plants.
- Organic mulches can improve soil structure and protect against surface compaction. The breakdown products of mulches help aggregate soil particles, increasing macropore space. Decomposition rates of mulches vary, however, as does the time required for the soil structure to improve. On wet soils, mulches may prevent soil drying, however, and this may contribute to root disease.
- Radial trenching may improve aeration status in localized areas of the root zone. This practice is a "soil-replacement"

strategy: poor-quality soil is replaced with higher-quality soil (fig. 5.33). For example, highly compacted soil is replaced with soil of a much lower bulk density. In some cases, a poor-quality soil is removed, amended, and then replaced. Soils with limited or reduced air-filled porosity may be improved by this technique, but the benefit is likely to be confined only to the volume of replacement soil.

Reducing soil moisture content

In many cases, high soil moisture content is directly responsible for inadequate aeration. Soil moisture should be maintained at levels that allow for adequate plant water supply and for adequate root zone aeration. Aeration can be improved by managing irrigation properly and improving drainage.

- Reduce water inputs: adjust the irrigation schedule to reduce application amount or frequency.
- Select species tolerant of high soil moisture conditions (fig. 5.34).
- Allow soil to dry between irrigations. The use of tensiometers may be useful to monitor soil moisture status.
- Improve surface drainage to avoid areas with standing water or ponding. Use drainage ditches or French drains to channel water out of low-lying areas. Contour the soil surface to direct water away from the affected area (see Harris, Clark, and Matheny 1999, table 7-2).
- Improve internal drainage (see Harris, Clark, and Matheny 1999, table 7-2).

Figure 5.32. Cultivation (A) and physical amendments (B) can increase the air-filled porosity of soils.

Figure 5.33. Soil replacement using a radial trenching technique can improve aeration in localized areas of the root zone. To improve aeration in the root zone of this California black oak *(Quercus kelloggii)*, soil with low air-filled porosity is being replaced using soil with a higher air-filled porosity.

- Where perched water tables are present, monitor the water level in the root zone by installing an observation well, such as a 2- to 4-inch-diameter plastic irrigation pipe.

Correcting surface barrier problems

- If a fill soil is present, conduct a root-crown excavation to determine the depth of the fill. Carefully remove soil from around the root crown. Remove all fill soil if practical; if not, monitor soil moisture levels in the fill and in the root zone, and maintain moisture levels adequate for water supply and aeration. For trees, construct a dry well (retaining wall) around the lower trunk.
- It is difficult to make changes in surface hardscape materials. Surface vents, air spaces, or barrier-free areas may improve soil aeration, although their impacts on ODR levels have not been quantified. Add organic mulch in areas where hardscape can be removed. Use species known to tolerate planting sites with extensive hardscape.

Figure 5.34. Selection of species tolerant of high soil moisture conditions is particularly important in irrigated landscapes that have poor drainage. In this planting, several species are declining or have been killed and removed due to an aeration deficit caused by excess irrigation and poor drainage. Note that some species do show tolerance.

Nutrient Deficiencies

Nutrient deficiencies, especially nitrogen, phosphorus, and potassium deficiencies, rarely occur in most woody landscape plants. Exceptions are palm trees, which often develop nitrogen, potassium, manganese, and magnesium deficiencies, and woody plants growing in soilless media in containers. When they do occur, nutrient deficiencies reduce shoot growth and leaf size and cause leaf chlorosis, necrosis, and dieback of plant parts. However, nutrient deficiencies cannot be reliably diagnosed on the basis of symptoms alone, as plant problems caused by numerous adverse growing conditions or pests produce similar symptoms. Leaf and/or soil samples are usually needed to confirm a deficiency. Also, nutrient availability depends in part on soil pH (see "Problems Related to pH," p. 117).

Nitrogen Deficiency

Symptoms

Nitrogen deficiency causes an overall decrease in vigor in all plants. Shoots are short and small in diameter, and leaves are small. In nitrogen-deficient broadleaf plants, older leaves become generally chlorotic (uniformly yellowish green), while young leaves may retain their typical green color (fig. 5.35). In fall, reddish leaf color is more pronounced, and leaf drop may occur earlier than normal. Species with compound leaves may have fewer leaflets. Flowering plants

Figure 5.35. Nitrogen deficiency symptoms in citrus (*Citrus* spp.) can be found in container plants (A) and in the field (B). Because nitrogen is mobile in the plant, chlorosis is more pronounced in older leaves. Nitrogen deficiency symptoms may be confused with symptoms caused by agents that restrict root growth and reduce nutrient uptake, including root pathogens, soil compaction, and poor drainage.

usually bloom normally, but bloom may be delayed.

In nitrogen-deficient conifers, needles are yellowish, short, and close together. In older conifers, the lower crown may be yellow, while the upper part remains green. Young nitrogen-deficient conifers may not develop side branches.

Nitrogen-deficient palm trees have decreased vigor and smaller, light green to yellow leaves. There is a gradation in color from old to young leaves, with the oldest leaves being most chlorotic. In severe cases, leaves are almost completely yellow or whitish, and growth ceases.

Occurrence

Except in palm trees, nitrogen deficiency is uncommon in established woody landscape plants. The deficiency may occur in newly planted trees, shrubs, and ground covers growing in very sandy or highly leached soils, or in plants growing in containers. Incorporating large amounts of non-decomposed organic amendments into the soil may also result in temporary nitrogen deficiencies in shallow-rooted plants. In palms, nitrogen deficiency commonly occurs in container-grown trees and in sandy, highly leached soils.

Look-Alike Disorders

Symptoms of nitrogen deficiency may be confused with symptoms caused by anything that restricts root growth and reduces nutrient uptake, including root diseases, insect pests, root pruning, soil compaction, adverse soil temperatures, aeration deficits, and poor drainage.

Diagnosis

Nitrogen deficiency may be diagnosed from visual symptoms, by leaf analysis, or by a combination of both methods. However, visual diagnosis alone is unreliable, since nitrogen deficiency symptoms can be caused by many things.

Plant Factors

- Examine the entire plant for overall health. Are mechanical injuries such as root cutting responsible for the reduced vigor and chlorotic leaves?
- Are the symptomatic plants young, vigorous, and growing in sandy or highly leached soil? Young plants with shallow, undeveloped root systems may be temporarily nitrogen deficient in sandy soils.
- Examine leaf symptoms carefully. Do symptoms appear on old or young leaves? Nitrogen deficiency symptoms occur first on the oldest leaves.

Site Factors

Visual examination

- Evaluate the soil. Is it wet and poorly drained? Plants growing in poorly aerated soil have reduced root growth and may be infected with root rot diseases, which can contribute to symptoms resembling nitrogen deficiency.
- Is there turfgrass growing beneath the tree? Normally, turfgrasses receive periodic applications of nitrogen fertilizer that would also be available to the tree.
- Are the plants growing in containers in soilless media? Such plants are more susceptible to nitrogen deficiency.

Leaf Analysis

A visual diagnosis should be combined with a leaf analysis. There are two ways to perform leaf analysis:

- Take leaf samples from trees that you suspect to be nitrogen deficient and from healthy trees of the same species growing nearby. Submit the samples to a laboratory (for more information on sampling, see chapter 2). Compare the results of the potentially nitrogen-deficient trees with the results of the healthy trees.
- Take leaf samples from trees that you suspect to be nitrogen deficient and compare the results with the leaf nitrogen ranges in table 5.3. The values in table 5.3 were obtained by sampling healthy, mature trees; they do not show the absolute minimum leaf nitrogen concentrations for healthy landscape trees. A value below the minimum given in the table does not

necessarily confirm nitrogen deficiency. A value above the minimum, however, very likely excludes nitrogen deficiency as a problem.

Soil Analysis

While soil analyses are useful for pH and salinity determinations, they are not reliable for determining the nitrogen needs or status of trees. The availability of nitrogen depends on soil condition and root volume, making a soil analysis report difficult to interpret. For example, the rate of nitrogen mineralization is affected by lack of soil oxygen, a common problem in frequently irrigated landscapes; also, the amount of available nitrogen in soil changes rapidly.

Remedies

Confirmed nitrogen deficiency may be corrected with soil applications of nitrogen-containing fertilizers.

TABLE 5.3. Total nitrogen (% total N) in leaves of selected landscape trees as given in published sources*

		Total nitrogen (% total N)		
Scientific name	Common name	Perry and Hickman 2001†	Mills and Jones 1996§	Kopinga and van den Burg 1995‡
Acer saccharinum	silver maple	2.0–3.4	2.3–2.6	1.9–2.7
Alnus rhombifolia	white alder	1.9–2.6	—	—
Betula pendula	European white birch	2.2–3.4	4.0–4.6	2.3–3.3
Cedrus deodara	deodar cedar	1.0–1.4	—	—
Celtis sinensis	Chinese hackberry	1.4–2.8	—	—
Cinnamomum camphora	camphor tree	1.2–2.0	—	—
Eucalyptus spp.	eucalyptus	1.8–2.1	1.1–1.2	—
Fraxinus angustifolia 'Raywood'	Raywood ash	2.1–2.9	—	—
Fraxinus velutina var. *glabra* 'Modesto'	Modesto ash	1.8–2.7	—	—
Ginkgo biloba	maidenhair tree	1.4–2.4	—	—
Gleditsia triacanthos	honey locust	2.3–3.1	2.4–4.0	2.0–2.5
Koelreuteria paniculata	goldenrain tree	1.9–3.5	2.5–2.8	—
Lagerstroemia indica	crape myrtle	1.1–3.5	1.6–2.1	—
Liriodendron tulipifera	tulip tree	1.2–2.8	1.9–4.3	2.6–3.0
Magnolia grandiflora	southern magnolia	1.0–3.5	—	—
Morus alba	white mulberry	2.0–3.6	1.2–2.4	—
Olea europaea	olive	1.3–1.9	—	—
Pistacia chinensis	Chinese pistache	1.6–3.0	2.1–2.8	—
Platanus acerifolia	London plane tree	1.4–2.6	2.0–2.7	2.0–2.6
Pyrus calleryana 'Bradford'	Bradford pear	1.1–1.9	1.6–2.5	—
Quercus ilex	holly oak	1.3–2.8	—	—
Quercus lobata	valley oak	2.1–2.9	—	—
Quercus suber	cork oak	1.5–2.2	—	—
Sapium sebiferum	Chinese tallow tree	1.7–2.7	—	—
Zelkova serrata	zelkova	1.8–2.8	2.3–3.1	—

Notes:

* This table does not give the absolute minimum leaf nitrogen concentrations of healthy landscape trees. A leaf sample value below the minimums shown in the table does not necessarily mean the tree is deficient in nitrogen.

† Samples taken from trees growing in landscapes.

§ Samples taken from trees growing in botanical gardens, field research plots, and field and container production nurseries.

‡ Source of samples not given.

Figure 5.36. Phosphorus deficiency symptoms in pear *(Pyrus communis)* leaves. Leaves are slightly smaller than normal and are distorted.

Phosphorus Deficiency

Symptoms

Visual symptoms of phosphorus deficiency are variable. In broadleaf plants, leaves may be dark green, especially when young. Leaf veins may be purple, especially on the underside of the leaves. Leaves may be sparse, slightly smaller than normal, and distorted, and they may drop earlier in fall (fig. 5.36). Older leaves may turn purple-bronze and have dead tips. Shoots are normal in length but smaller in diameter. Phosphorus-deficient flowering plants produce fewer flowers.

In phosphorus-deficient conifers, the needles on older trees are dull blue-green or gray-green. As the deficiency worsens, the needles die, starting low in the tree and progressing upward. Few or no secondary needles may grow. In seedlings, needles turn purple, starting at the tips of lower needles and progressing inward and upward.

Occurrence

Phosphorus deficiency is rare in landscape plants, as few soils are phosphorus deficient. Young or shallow-rooted plants may become phosphorus deficient where topsoil has been removed.

Look-Alike Disorders

Certain herbicides cause leaf distortion and a red to purple color change in leaves, followed by marginal necrosis. Symptoms occur in older leaves first, as with phosphorus deficiency.

Diagnosis

Phosphorus deficiency may be diagnosed by a combination of leaf analysis and visual symptoms.

- Has the soil been graded? Phosphorus does not move readily in the soil and tends to remain concentrated in the surface soil. Young or shallow-rooted plants may become phosphorus deficient where topsoil has been removed.
- Take leaf samples from suspected phosphorus-deficient trees and healthy trees of the same species growing nearby (for leaf sampling information, see chapter 2). Compare the results of the two analyses.
- A conifer or broadleaf plant may be phosphorus deficient if the level in the current season's leaves is below 0.1 percent.

Remedies

Confirmed phosphorus deficiency may be corrected with soil applications of phosphorus-containing fertilizers.

Potassium Deficiency

Symptoms

The leaves of potassium-deficient broadleaf plants develop marginal and interveinal chlorosis, followed by necrosis. The necrosis progresses inward until the entire leaf blade is scorched. The most recently matured leaves are affected first. With time, the symptoms become more pronounced on the older leaves. Leaves may crinkle and roll upward. The tips of the current season's shoots die back late in the season. Fewer flower buds form.

In potassium-deficient conifers, the oldest foliage turns dark blue-green, progressing to yellow and reddish brown. The needles are often stunted, and the tips of needles become necrotic.

Symptoms of potassium deficiency in palms vary among species but always appear first on the oldest leaves (figs. 5.37–39). The most common symptom in palms is yellow stippling on leaves, which progress to the younger leaves as the deficiency worsens. The yellow spots appear translucent when viewed from below. In some species, the symptoms are marginal or tip necrosis of the older leaves, with little or no yellow spotting. The most severely affected leaves or leaflets are completely necrotic and withered. Leaf midribs may be orange, rather than green. In date palms (*Phoenix* spp.), the tips of older leaves develop a dull orange-brown discoloration. Most palms are susceptible to potassium deficiency, but it is most severe in royal (*Roystonea* spp.), queen (*Syagrus romanzoffianum*), coconut (*Cocos nucifera*), date (*Phoenix* spp.), and areca (*Areca* spp.) palms.

Figure 5.37. Yellow stippling on leaves indicates potassium deficiency symptoms in queen palm *(Arecastrum romanzoffianum)*. Symptoms of potassium deficiency in palms vary between species but always appear first on the oldest leaves.

Occurrence

Potassium deficiency is extremely rare in broadleaf and coniferous landscape plants, but relatively common in palm trees. In palms, the deficiency develops readily in sandy, rapidly drained soils, especially

Figure 5.38. Potassium deficiency symptoms in Oaxaca palmetto *(Arecastrum romanzoffianum)*. The most common symptom in palms is yellow stippling on the older leaves that progresses to younger leaves as the deficiency worsens.

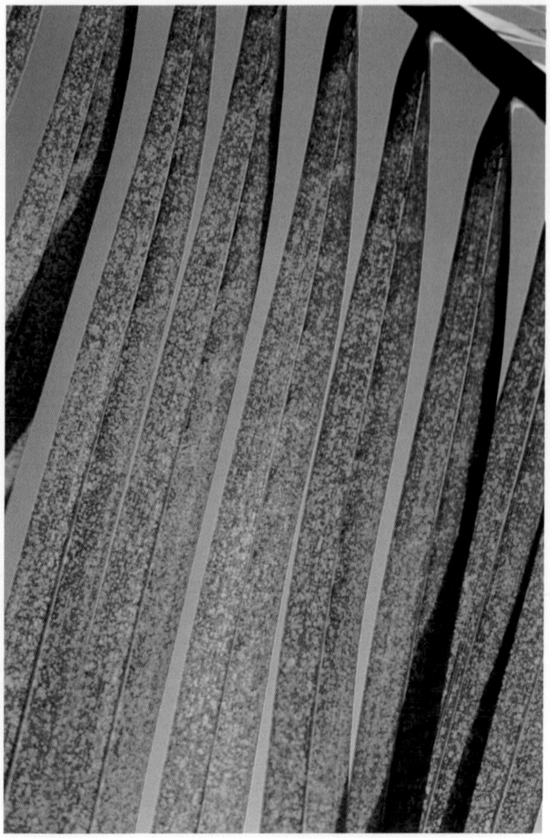

Figure 5.39. Potassium deficiency symptoms in pindo palm *(Butia capitata)*. In palms, potassium deficiency develops readily in sandy, rapidly drained soils, especially where frequent irrigation is being applied, or in palms growing in container media.

where frequent irrigation is being applied, or in container media. It is relatively uncommon in palms growing in clay soils.

Look-Alike Disorders

Early symptoms of potassium deficiency are often confused with leaf spot diseases or the yellow and brown spots caused by sucking insects. Certain preemergence herbicides cause marginal and interveinal chlorosis followed by necrosis. Preemergence systemic herbicide symptoms occur in older leaves first, as with potassium deficiency.

Diagnosis

Potassium deficiency is diagnosed primarily from visual symptoms. Leaf analysis may be unreliable, as potassium is mobile in plants and may be leached from leaves with rain or sprinkler irrigation.

- Has the soil been graded? The highest concentration of potassium in soil usually occurs near the surface. Young or shallow-rooted plants may become potassium deficient where topsoil has been removed.
- If the deficiency is suspected in a palm, examine the soil. Sandy, rapidly drained soils are most susceptible to potassium deficiency.

Figure 5.40. Iron deficiency symptoms in sweetgum *(Liquidambar styraciflua)*; non-deficient leaf on right. Because iron is immobile in plants, older basal leaves remain green as young leaves become chlorotic between veins. Various soil conditions are usually responsible for making iron unavailable to plants, including a soil pH above 7.0 and cold, wet, or poorly drained soils.

Sensitive Species

Many palm species are sensitive to potassium deficiency, including royal (*Roystonea* spp.), queen (*Arecastrum romanzoffianum*), coconut (*Cocos nucifera*), date (*Phoenix* spp.), and areca (*Areca* spp.) palms.

Remedies

Confirmed potassium deficiency in palms may be corrected with soil applications of potassium-containing fertilizers. On very sandy soils, slow-release forms of potassium fertilizers should be applied to prevent rapid loss due to leaching. Use organic mulch to cover soil around plants, as potassium leaches readily from organic matter. In palm trees, symptomatic leaves do not recover and must be replaced by new growth. However, do not remove symptomatic leaves, as this may cause the deficiency to worsen.

Iron Deficiency

Symptoms

Iron deficiency, or iron chlorosis, is the yellowing of leaves due to a lack of iron. In iron-deficient broadleaf plants, the areas between the veins of young leaves are chlorotic, with distinct narrow green veins (fig. 5.40). Because iron is immobile in plants, the older basal leaves remain green as young leaves become chlorotic. With extreme iron deficiency, young leaves are small, almost white, and may have necrotic margins and tips. Shoots are normal in length but small in diameter. Twigs die back and defoliate as the deficiency increases in severity (fig. 5.41).

In iron-deficient conifers the older needles and the lower crown remain green, while new needles are stunted and chlorotic. The new leaves of iron-deficient palm trees are uniformly chlorotic, especially in poorly aerated soil or when the tree has been planted too deep.

Occurrence

In many areas, iron chlorosis is the most common nutrient deficiency seen in landscape plants. The concentration of iron in soil is usually adequate for plant growth, but various soil conditions can make the element unavailable to plants. For example, the availability of soil iron decreases as soil pH rises above 7.0, and it decreases in cold, wet, or poorly drained soils. Symptoms of iron deficiency are variable within trees, between adjacent trees, and between species. Iron deficiency in palm trees is uncommon; it can appear in palms growing in poorly aerated soils or those that have been planted too deeply. In soils above pH 7.0, species adapted to acidic soils ("acid-loving" plants) are usually affected by iron deficiency symptoms.

Look-Alike Disorders

Symptoms of iron deficiency closely resemble those of manganese deficiency or damage by soil-applied preemergence systemic herbicides. However, iron deficiency occurs on new leaves, and preemergence herbicide injury occurs primarily on old leaves. Soil compaction, root injury caused by construction, and poor drainage can aggravate iron deficiency symptoms or cause symptoms similar to iron chlorosis.

Figure 5.41. Advanced iron deficiency symptoms in sweetgum *(Liquidambar styraciflua)*. The twigs of this tree are dying and defoliating as the deficiency increases in severity.

Diagnosis

Iron deficiency may be diagnosed by a combination of visual symptoms and soil analysis.

Plant Factors

- Identify the species. Certain acid-loving species are susceptible to iron deficiency in alkaline soils.
- Examine leaf symptoms carefully. Do symptoms appear on old or young leaves? Because iron is not mobile in the plant, iron deficiency symptoms appear first on the youngest leaves.
- The rapid growth of vigorous young trees is often temporarily iron deficient in spring.
- Iron deficiency symptoms may occur on few plants within a group of the same species in the same landscape. Or, individual branches within a tree or shrub may be affected.

Soil Factors

- Analyze the soil for pH. As soil pH increases above 7.0, iron availability decreases.
- Evaluate the soil. Is it wet and poorly drained? Plants growing in soils low in oxygen are prone to iron deficiency symptoms, especially in spring when soil temperatures are low.

Sensitive Species

Plants adapted to acidic soils (see table 5.15, p. 122) are often iron deficient in alkaline soils.

Remedies

Confirmed iron deficiency may be corrected with soil applications of iron-containing fertilizers. For a more rapid correction, iron chelates may be applied to the soil or foliage, but the effect of iron chelates is relatively short-lived. Trunk injections of iron

Figure 5.42. Manganese deficiency symptoms in red maple *(Acer rubrum)*. The new leaves become yellow, with wide green bands along the veins.

also temporarily correct iron chlorosis, but they injure the trunk. When iron deficiency is caused by adverse soil conditions, long-term correction is achieved by adding soil amendments to lower pH (soil acidification) or by cultural practices such as avoiding overwatering and improving soil aeration.

Manganese Deficiency

Symptoms

In manganese-deficient broadleaf plants, new leaves become yellow with wide green bands along the veins (fig. 5.42). As the element concentration decreases, interveinal necrotic spots develop. In some plants, leaf margins may be wavy, crinkled, or curled. Shoot growth may be reduced.

Symptoms of manganese deficiency in conifers are very similar to those of iron deficiency. New growth is stunted and chlorotic, and older needles and the lower crown remain green.

The new leaves of manganese-deficient palms are uniformly chlorotic with interveinal necrotic streaks and are smaller than old leaves. As the deficiency worsens, newly emerging leaves are necrotic, distorted, and withered. Older leaves may be completely withered and distorted ("frizzletop"), scorched, and greatly reduced in size. Affected queen *(Arecastrum romanzoffianum)* and pygmy date *(Phoenix roebelenii)* palms develop a flat-headed appearance caused by leaf stunting. In severely deficient palms, growth ceases, and emerging leaves consist only of necrotic and stunted petioles.

Occurrence

The total amount of manganese in soil is usually adequate for plant growth, yet soil factors often make the element unavailable to plants. Although deficient in landscape plants less often than iron, manganese is usually deficient under similar conditions and in a variety of landscape plants. The availability of manganese to plants decreases as soil pH rises above 6.5, and it decreases in poorly drained soils that are high in organic matter. The deficiency often occurs when soils are alternately well drained and waterlogged. Symptoms are also more likely to occur in plants suffering from drought. Manganese deficiency in palm trees is very common in alkaline soils, but it can also occur in poorly drained soils or where soil temperatures are cool.

Look-Alike Disorders

Manganese deficiency symptoms closely resemble those of iron deficiency and may also be confused with symptoms caused by soil-applied preemergence herbicides; however, manganese deficiency occurs on new leaves, and preemergence herbicide injury occurs primarily on old leaves first.

Diagnosis

Manganese deficiency may be diagnosed by a combination of visual symptoms and soil analysis.

Plant Factors

- Identify the species. Certain acid-loving species are susceptible to manganese deficiency in alkaline soils.
- Examine leaf symptoms carefully. Do symptoms appear on old or young leaves? Because manganese is not mobile in the plant, deficiency symptoms occur first on the youngest leaves.

Soil Factors

- Analyze the soil to determine pH. As soil pH increases above 6.5, the availability of manganese decreases.
- Evaluate the soil. Is it wet and poorly drained? Plants growing in poorly aerated soil are prone to manganese deficiency.

Sensitive and Resistant Species

Plants adapted to acidic soils (see table 5.15, p. 122), such as azaleas and gardenias, are often manganese deficient in alkaline soils. In palms, manganese deficiency is most common in queen palm *(Arecastrum romanzoffianum)*, pygmy date palm *(Phoenix roebelenii)*, and sago palm *(Cycas revoluta)*, while it is uncommon in *Washingtonia* spp. palms.

Remedies

Confirmed manganese deficiency may be corrected with soil and foliar applications of manganese-containing fertilizers. The symptoms are usually alleviated by adding soil amendments to lower pH (soil acidification) or by cultural practices such as avoiding overwatering and improving soil drainage.

Figure 5.43. Zinc deficiency symptoms in almond *(Prunus dulcis)*. This deficiency causes extremely shortened internodes, resulting in tufts of leaves (rosettes, or witches' broom) at the tips of small-diameter shoots.

Zinc Deficiency

Symptoms

In zinc-deficient broadleaf plants, leaves are uniformly yellow and sometimes mottled with necrotic spots. Leaves are small ("little-leaf" symptom), very narrow, and pointed. The deficiency causes shortened internodes, resulting in tufts of leaves (rosettes, or witches' broom) at the tips of small-diameter shoots (fig. 5.43). Older leaves eventually drop.

In zinc-deficient conifers, needles are stunted and chlorotic. Branches may also be stunted and may die back. Trees may lose all but first- or second-year needles.

Occurrence

The amount of zinc in soil is usually adequate for plant growth, yet soil factors often make the element unavailable to plants. The availability of zinc to plants decreases as soil pH rises (becomes more alkaline). The symptoms are variable within the tree and between adjacent trees. Since the level of zinc is highest in surface soils, grading and other practices that reduce rooting in surface soil may increase zinc deficiency. Zinc deficiency is rare in palms.

Look-Alike Disorders

Zinc deficiency symptoms may be confused with symptoms of glyphosate (systemic herbicide) injury.

Diagnosis

Zinc deficiency may be diagnosed by a combination of visual symptoms and soil analysis.

Plant Factors

- Examine leaf symptoms carefully. Tufts (rosettes) of small, yellow leaves at the tips of shoots are distinctive symptoms of zinc deficiency.

Figure 5.44. Magnesium deficiency symptoms in pygmy date palm *(Phoenix roebelenii).* As typical for magnesium deficiency in palms, the ends of these fronds are bright yellow.

Figure 5.45. Magnesium deficiency symptoms in Senegal date palm *(Phoenix reclinata).* Date palms (*Phoenix* spp.) are most susceptible to this deficiency.

Soil Factors

- Analyze the soil to determine pH. As soil pH increases above 6.5, the availability of zinc decreases. Zinc deficiency is also aggravated in soils high in phosphorus.
- Has the soil been graded? Since the level of zinc is highest in surface soils, grading and other practices that reduce rooting in surface soil, such as frequent shallow cultivation, may increase zinc deficiency.

Remedies

Confirmed zinc deficiency may be corrected with applications of zinc-containing fertilizers. For a more rapid correction, zinc chelates may be applied to the soil or foliage, but the effect of zinc chelates is relatively short-lived. Symptoms may be alleviated by adding soil amendments to lower pH (soil acidification).

Magnesium Deficiency

Symptoms

In palms, the ends of leaves become bright yellow. Individual fronds have a band of yellow around the outer perimeter, while the midrib and portions of leaves near the midrib remain green. The lower fronds senesce prematurely (figs. 5.44–45).

Occurrence

Magnesium deficiency is common on palm trees growing in the Southwest. Date palms (*Phoenix* spp.) are most susceptible. The deficiency is rare in broadleaf and coniferous landscape plants.

Look-Alike Disorders

None.

Remedies

Magnesium deficiency in palms may be corrected with soil applications of magnesium sulfate. Symptomatic leaves do not recover and must be replaced by new growth. Do not remove symptomatic leaves until the leaves have turned brown.

Salinity (Salt Injury) and Specific Ion Toxicty

Soils contain a mixture of water-soluble salts that are necessary for plant growth and function. When present in high concentrations, however, salts can injure sensitive plants. The salts that cause plant damage are primarily chlorides, sulfates, and, less frequently, nitrates of calcium, magnesium, sodium, and potassium. Salts may be formed during the weathering of rocks and minerals or they may be added to the soil through rainfall, irrigation, fertilization, incorporation of saline soil amendments, or application of deicing salts in snow-prone areas. Soils containing high concentrations of salt are called *saline soils*.

Soils that are high in exchangeable sodium are called *sodic* or *alkali* soils. Sodic soils have a high concentration of sodium relative to calcium and magnesium (sodium is more than 15 percent of the exchange capacity). They are problematic for plants in two ways. Sodium causes medium- or fine-textured soil to lose its aggregated structure, becoming impervious to air and water. Sodium can also be toxic to sensitive plants. The pH of sodic soils usually exceeds 8.5. Soils that contain salts and high levels of exchangeable sodium are called *saline-sodic*.

Figure 5.46. These *Hebe* 'Coed' plants demonstrate salinity effects. Plant on left shows normal growth where salinity was low (1 mmhos/cm); center plant, reduced growth at 3 mmhos/cm; and right, leaf necrosis and defoliation where salinity was high (6 mmhos/cm).

Plants absorb salts through the foliage and the roots. The most common sources of foliar-applied salts are ocean spray, sprinkler irrigation with poor-quality water, and, in cold areas, splash from road deicing salts.

Individual elements or compounds may cause injury. Chloride, boron, and ammonium are injurious when their concentrations reach a critical level in sensitive plants. This is known as *specific ion toxicity*.

Symptoms

Root-Absorbed Salts

Salt toxicity is first expressed as stunting of growth and yellowing of foliage. In broadleaf species, leaf necrosis and defoliation usually follows (fig. 5.46). Typically, the symptoms are usually most severe on the edges and tip of older leaves where the greatest salt accumulation typically occurs and less severe on new foliage (fig. 5.47). For conifers, needles turn yellow, then brown from the tip downward and defoliate (fig. 5.48). In severe cases, plants are killed. The degree of the symptoms depends on the sensitivity of the plant to salts and the concentration of accumulated salts in the soil.

Foliar-Applied Salts

Plants absorb salts through their foliage and roots. Salt spray damage is generally confined to the portion of the canopy exposed to wind-driven spray. Along the coast, for instance, the greatest damage would be on the windward side of the plant. In the case of road-applied salts, the symptoms would be most severe on the side of the plant facing the road and to the height of the splash generated by traffic (fig. 5.49). Conifers are likely to show foliar damage from deicing salts in the winter months. Needles turn reddish brown from the tips downward, then become darker brown and brittle, often breaking off in the wind. Foliage of broadleaf plants may show marginal necrosis,

Figure 5.47. On broadleaf plants, salt toxicity symptoms include leaf chlorosis and necrosis of tips and margins, with the symptoms being most severe on the older leaves, as in this European hackberry *(Celtis australis)*.

Figure 5.48. On conifers, excessive salts cause necrosis from the tip down the needle; symptoms are most severe on older growth.

defoliation, premature fall coloration, and delayed spring leafout. For conifers and broadleaf species, bud, twig, branch, and whole plant death may occur. In deciduous broadleaf species, terminal bud or twig death may lead to a witches' broom growth the following spring (Laemmlen 1980). The severity of symptoms depends on the sensitivity of the species, the length of the exposure, and the concentration of salts in the spray.

Sodic (Alkali) Soils

Plants growing in sodic soils may sustain injury ranging from stunted growth and chlorosis to necrosis and death. Presence of a white or black crust on the soil surface is one indication of an alkali condition. Excessive sodium may be suspected if water penetration becomes increasingly slower.

Specific Ion Toxicity

The toxicity symptoms of specific ions are often difficult to distinguish from each other. Leaf chlorosis and tip and marginal necrosis are typical.

Chloride toxicity symptoms include reduced leaf size and slower growth rate, necrosis of leaf tips or margins, bronzing, premature yellowing and abscission of leaves, and chlorosis (Chapman 1965) (fig. 5.50).

In the early stages, sodium toxicity appears as foliar mottling or interveinal chlorosis (fig. 5.51). Damage progresses to necrotic leaf tips, margins, or interveinal areas.

Early stages of boron toxicity are generally characterized by yellowing of the leaf tip, followed by progressive chlorosis and necrosis of margins and between lateral veins (Chapman 1965). Necrosis associated with boron is often black and may appear as small spots near the leaf margin (fig. 5.52). In acute stages, premature leaf drop occurs. On conifers, needles die from the tip downward; this is most severe on the older foliage (fig. 5.53).

Ammonium toxicity symptoms include reduced growth rate, chlorosis, development of small necrotic spots on leaves, and blackening of the stem (Bunt 1976) (fig. 5.54).

Occurrence

Soil Salinity

Toxicities may occur when plant roots absorb salts from the soil. Salinity is a problem in many arid, coastal, and snow-prone regions. Salts may be present from the soil parent material or may accumulate in the soil from irrigation water, high groundwater tables, or deicing salts (Harris, Clark, and Matheny 1999). Saline soils are found most often in areas with precipitation-to-evaporation ratios of 0.75 or less, and in low, flat areas with high water tables that

Figure 5.49. Foliar-applied salts cause a red tip burn on conifer needles (A). The pattern of damage conforms to the pattern of spray and splatter from traffic that disperses road-applied salts (B).

Figure 5.50. Chloride toxicity symptoms are reduced leaf size, chlorosis, and necrosis of leaf tips and margins on sweetgum *(Liquidambar styraciflua)* (A) and European hackberry *(Celtis australis)* (B).

Figure 5.51. Sodium toxicity symptoms on star jasmine *(Trachelospermum jasminoides)* first appear as foliar mottling or interveinal chlorosis.

Figure 5.52. Boron toxicity causes dark, necrotic areas along the leaf margins in white mulberry *(Morus alba)*.

Figure 5.53. Severe boron toxicity symptoms on redwood *(Sequoia sempervirens)* causes needle death. Symptoms appear first and are more severe on older foliage.

may be subject to seepage from higher elevations (Brady and Weil 1996).

Road salt is used in some snowy areas to improve road conditions. Sodium chloride (NaCl) and calcium chloride ($CaCl_2$) are the salts commonly used. Low-lying areas adjacent to treated roads typically have the highest salt concentrations. Deicing salts are less of a problem in the root zone in wet springs than in dry springs because rain leaches salts below the root zone (Michigan State Univ. Ext. 1996).

Saline soils may be found in unexpected places. Saline soils imported into a site can introduce salts into an area that is normally nonsaline. Also, grade cuts may expose underlying saline soils. Consider changes to the natural topography when assessing soil characteristics.

Figure 5.54. Ammonium toxicity caused stunted growth and necrotic spots on chrysanthemum *(Chrysanthemum morifolum)* (asymptomatic plant on left).

Irrigation with saline water increases soil salinity over time. If irrigation water does not penetrate the soil deeply, most of the salt contained in that water will remain within the plant root zone and damage sensitive plants. The higher the salinity of the water, the more frequent the irrigation, and the poorer the drainage, the greater the potential for saline soil to develop. Recycled water, water from shallow wells, and water from rivers that pass through arid agricultural regions may be saline.

Excessive application of fertilizer can cause salt toxicity. In addition, the formulation of the fertilizer affects the contribution to soil salinity. For instance, potassium chloride (KCl, muriate of potash) has a higher salt hazard than potassium sulfate (K_2SO_4) (table 5.4). Commonly used soil amendments such as animal manures, mushroom compost, and sewage sludge may be very high in salts (table 5.5). When incorporated into soils, these materials may damage plants if soil salinity exceeds plant tolerance.

Foliar-Applied Salts

Toxicity also can occur from foliar-applied salts. In fact, plants are usually damaged at lower salt concentrations when salts are applied to the foliage than to the soil. Salts may be deposited on the foliage through wind-blown ocean spray or splash of road deicing salts. Ocean spray damage to sensitive species has been observed 12 to 18 miles (about 20 to 30 km) inland. Plant damage has occurred 300 yards (275 m) downwind of heavily traveled highways in New England that have been treated with deicing salts (Pennsylvania State Univ. 1987), although damage is most severe within 60 feet (18 m) of the road (Johnson and Sucoff 1995). Sprinkler irrigation with poor-quality water can lead to damage of sensitive species.

Sodic Soils

Sodic soils usually develop in low-lying areas with high water tables or areas in which saline and alkaline waters have ponded for many years. Irrigating with water high in sodium also causes sodic soils.

Look-Alike Disorders

Salt damage can appear similar to many other disorders, including mineral deficiencies, water deficit, herbicide toxicity, wind burn (desiccation), acute air pollution injury, and high light exposure.

TABLE 5.4. Effect of selected fertilizers on soil salinity

Fertilizer	% Element	Salt index*	Relative salinity†
ammonium nitrate	35 N	104.7	49.4
ammonium sulphate	21 N	69.0	53.7
calcium nitrate	11.9 N	52.5	30.1
diammonium phosphate	21 N, 23 P	34.2	12.7
monoammonium phosphate	12 N, 27 P	29.9	12.7
potassium chloride	49.8 N	116.3	38.5
potassium nitrate	13 N, 38 K	73.6	23.6
potassium sulfate	45 K	46.1	17.0
sodium nitrate	15.5 N	100.0	100.0
superphosphate (single)	7.8 P	7.8	16.5
superphosphate (triple)	19.6 P	10.1	8.5
urea	46 N	75.4	26.7

Source: Bunt 1976.
Notes:
*Salt index is calculated from the increase in osmotic pressure of equal weights of fertilizers, relative to the effect of $NaNO_3$
†Relative salinity is calculated from the increase in osmotic pressure per unit of plant nutrient, relative to the effect of $NaNO_3$.

Figure 5.55. These sugar gums *(Eucalyptus cladocalyx)* were damaged by high soil salinity at a depth of 4 feet (1.2 m). Salinity in the surface was within a tolerable range.

Conditions under Which Salt or Specific Ion Toxicity May Occur

Specific ion toxicity can occur under the same conditions as described for salinity. Most toxicities result from naturally high concentrations of ions in the soil and water. Plants absorb sodium and chloride through the foliage and roots.

Irrigating with softened water that displaces calcium for sodium or discharging softened water into leach fields or septic systems may damage plants. Soil drainage problems may develop if sodium is not adequately balanced by calcium and magnesium (see "Sodic Soils," p. 85). Sodium also may be high if sodium chloride (NaCl) is used as a deicing salt.

Chloride toxicity can occur when sodium chloride and calcium chloride ($CaCl_2$) are used as deicing salts. Sensitive plants growing adjacent to chlorinated swimming pools may be damaged by splashed water and filter backwash. Heavy or repeated fertilization containing chloride (e.g., muriate of potash, KCl) may also cause toxicity symptoms.

Boron occurs in soils derived from marine sediments or parent material rich in boron-containing minerals, arid soils, and soils derived from geologically young deposits. Soil management practices that produce boron toxicity include acidification of certain neutral or alkaline soils, irrigation with high-boron-content water, and application of some sewage effluent wastes (Chapman 1965). Soils that have been treated with borate-containing herbicides may be toxic.

Ammonium toxicity can occur from excess application of ammonium fertilizers (see table 5.4) or incorporation of soil amendments high in ammonium .

Diagnosis

Soil salinity can be diagnosed by collecting a soil sample from the root zone and measuring the electrical conductivity (EC_e) of a saturated paste. Collect samples from the appropriate depth; determine where the roots are located when deciding depth of sampling. Salts often accumulate in the top ½ inch (1.3 cm) of soil. Tests on samples from the surface would be representative only of plant growing conditions if the area were a seedbed. Where plants are deeply rooted, samples must be deep. In diagnosing sugar gum tree dieback on a low-lying area near the San Francisco Bay, salinity was found to increase with soil depth and reached toxic levels only at 4 feet (1.2 m) (fig. 5.55).

EC_e is usually expressed in millimhos per centimeter (mmhos/cm) or decisiemens per meter (dS/m); 1 dS/m equals 1 mmhos/cm. The lower the salt content of the soil, the lower the EC_e and the less potential for plant injury. Soils are generally considered saline when EC_e exceeds 3 mmhos/cm. At that concentration, sensitive plants may be damaged. Woody plants considered to be tolerant may not show damage until the EC_e reaches 10 mmhos/cm.

The salinity of water is expressed as total dissolved solids (TDS) and EC_w. For most plants, TDS should be below 1,000 ppm and EC_w should be below 2 mmhos/cm.

While salinity expresses the total salt content, it does not adequately identify potential toxicities from specific ions such as chloride, sodium, boron, and ammonium. Boron in particular must be evaluated independently; it is toxic in such low concentrations (less than 1 ppm) that its presence will not be reflected in the normal salinity measurement.

Take water and leaf samples to diagnose specific ion toxicity and better understand the salt source and possible treatments (for interpretive guidelines, see table 5.6). Collect older, exposed leaves for tissue testing: salts accumulate in plant tissues with age, so older leaves and those exposed to

TABLE 5.5. Salt content of commercially available organic soil amendments, ranked by EC_e

Product	Number of brands tested	EC_e (mmhos/cm)*
chicken manures	2	17.0–23.0
steer manures	3	12.0–14.0
mushroom composts	1	13.0
"organic" composts	4	2.9–7.9
planting mixes	2	0.6–2.0
peat	1	0.3
redwood composts	2	0.2–1.8

Source: Hesketh and Hickman 1984.
Note: *EC_e from 0 to 2.0 mmhos/cm is considered safe for most plants.

TABLE 5.6. Guidelines for interpreting test results for salts in soil, water, and plant tissue

	Generally safe	Slight to moderate	Severe
Soil analyses			
salinity (EC_e), mmhos/cm*	0.5–2.0	2.0–4.0	>4.0
sodium adsorption ratio (SAR)	<6	7–9	>9
sodium, mg/l		>230	
boron, mg/l	0.1–0.5	1–5	>5
chloride, mg/l	<100	100–200	>250
ammonium, mg/l	0–25	>25	
Water analyses*			
total dissolved solids (TDS), mg/l	<450	450–2,000	>2,000
salinity (EC_w), mmhos/cm†	<0.7	0.7–3.0	>3.0
boron, mg/l	<0.5	0.5–1.0	>1.0
chloride			
surface irrigation, mg/l	<140	140–300	>350
sprinkler irrigation, mg/l	<100	>100	
sodium			
surface irrigation (SAR)	<3	3–9	>9
sprinkler irrigation, mg/l	<70	>70	
Tissue analyses			
boron, ppm			>200
sodium, %	0.1–0.3		>5
chloride, %	0.2–0.4		>1

Notes: *1 mmhos/cm = 1 dS/m.
†*Source*: Pettygrove and Asano 1985.

the atmosphere typically have higher concentrations of salts than younger or interior leaves. When interpreting data, compare it with analyses of asymptomatic foliage of the same age, position, and species.

Test for sodic conditions by measuring the calcium, magnesium, and sodium in the soil and calculating the sodium adsorption ratio (SAR) or exchangeable sodium percentage (ESP). The SAR should be less than 6.0; the ESP, below 15.

For information on how to collect soil, water and plant tissue samples, see chapter 2.

Sensitive and Tolerant Species

Plants vary considerably in their tolerance to salinity; knowing the EC_e of the soil and the tolerance of the species are key to diagnosing damage due to salt toxicity (see tables 5.7–10) Furthermore, plants that are tolerant to soil salinity may be sensitive to salt spray.

Plants have different tolerances to specific ions. For instance, a plant may be relatively tolerant to boron but highly sensitive to chloride. Furthermore, genetic variations in sensitivity are found within species. Limited information is available to help develop lists of plants sensitive to specific ions. Salt tolerance lists are probably adequate to assess chloride sensitivity. A list of plants sensitive and tolerant to boron is provided in table 5.9, p. 111; on plant tolerance to sodic soil, see table 5.15, p. 122.

Remedies

Saline soils cannot be remedied by any chemical amendment, conditioners, or fertilizer (Cardon and Mortvedt 1999). The key to reclamation is leaching the salts below the plant root zone through drainage and good-quality water. Problems with drainage must be corrected before leaching treatments are applied (see Harris, Clark, and Matheny 1999). In cases where irrigation water is more saline than tolerable for sensitive plants, leaching will not lower the salinity below the injurious level (fig. 5.56, p. 115).

TABLE 5.7. Salt and boron tolerance of selected landscape plants

Scientific name	Common name	Plant type	Salinity tolerance	Boron tolerance	Reference*
Abelia × *grandiflora*	glossy abelia	shrub	low	moderate	Francois 1980, 1982; Skimina 1980
Abelia × *grandiflora* 'Edward Goucher'	Edward Goucher abelia	shrub	low	—	Wu et al. 1999
Abies balsamea	balsam fir	tree	low	—	Dirr 1997
Acacia farnesiana	sweet acacia	tree	moderate	—	Cal Poly n.d.
Acacia greggii	catclaw acacia	tree	high	—	Cal Poly n.d.
Acacia longifolia	Sydney golden wattle	shrub	low	—	Farnham, Ayers, and Hasek 1985
Acacia melanoxylon†	blackwood acacia	tree	high	—	Roewekamp 1941
Acacia melanoxylon†	blackwood acacia	tree	moderate	moderate	Bernstein, Francois, and Clark 1972; Van Arsdel 1980
Acacia redolens	prostrate acacia	groundcover	high	—	Wu et al. 1999
Acacia spp.	acacia	tree	moderate	high	Cal Poly n.d.; San Diego 1966
Acalypha wilkesiana	copper leaf	shrub	low	—	Cal Poly n.d.
Acanthus mollis 'Oak Leaf'	bear's breech	shrub	low	—	Skimina 1980
Acer buergerianum	trident maple	tree	high	—	Cornell Coop. Ext. 1999
Acer campestre	hedge maple	tree	high	—	Cornell Coop. Ext. 1999
Acer macrophyllum	bigleaf maple	tree	low	—	Donaldson, Hasey, and Davis 1983
Acer platanoides	Norway maple	tree	moderate	—	Branson and Davis 1965; Dirr 1978
Acer pseudoplatanus†	sycamore maple	tree	high	—	Cornell Coop. Ext. 1999
Acer pseudoplatanus†	sycamore maple	tree	low	—	Butin 1995
Acer rubrum	red maple	tree	low	—	Cornell Coop. Ext. 1999
Acer saccharinum	silver maple	tree	low	low	Bernstein, Francois, and Clark 1972; Van Arsdel 1980
Achillea spp.	yarrow	herbaceous	moderate	—	Glattstein 1989
Aesculus carnea	red horsechestnut	tree	—	low	Questa 1987
Aesculus spp.	horsechestnut	tree	low	—	Butin 1995
Agapanthus orientalis	lily-of-the-Nile	herbaceous	moderate	—	Skimina 1980
Agapanthus umbellatus	lily-of-the-Nile	herbaceous	moderate	—	Skimina 1980
Agave attenuata	thin-leaved agave	shrub	moderate	—	Farnham, Ayers, and Hasek 1985
Agave spp.	agave	shrub	high	—	Van Arsdel 1980
Ageratum spp.	ageratum	herbaceous	moderate	—	Morris and Devitt 1990
Agrostis tenuis	colonial bentgrass	turfgrass	low	—	Harivandi 1988
Ailanthus altissima	tree-of-heaven	tree	high	—	Butin 1995
Albizia julibrissin	silk tree	tree	low	—	Questa 1987; Wu et al. 1999
Albizia lophantha (syn. *A. distachya*)	plume albizia	tree	high	—	Cal Poly n.d.
Allamanda cathartica	allamanda	vine	moderate	—	Cal Poly n.d.
Alnus incana	speckled alder	tree	low	—	Dirr 1997
Alnus rhombifolia	white alder	tree	low	low	Bernstein, Francois, and Clark 1972; Van Arsdel 1980
Aloe arborescens	tree aloe	shrub	high	—	Cal Poly n.d.

Scientific name	Common name	Plant type	Salinity tolerance	Boron tolerance	Reference*
Aloe vera	medicinal aloe	herbaceous	high	—	Cal Poly n.d.
Alyssum saxatile	alyssum	herbaceous	low	—	—
Antirrhinum majus	snapdragon	herbaceous	low	—	—
Araucaria araucana	monkey puzzle	tree	low	—	Van Arsdel 1980
Araucaria heterophylla†	Norfolk Island pine	tree	high	—	Farnham, Ayers, and Hasek 1985; Skimina 1980
Araucaria heterophylla†	Norfolk Island pine	tree	moderate	—	Van Arsdel 1980
Arbutus unedo†	strawberry tree	tree	moderate	—	Skimina 1980
Arbutus unedo	strawberry tree	tree	high	high	Cal Poly n.d.; Roewekamp 1941; Questa 1987
Arbutus unedo 'Compacta'	compact strawberry tree	shrub	low	—	Francois and Clark 1978; Skimina 1980
Arctotheca calendula	Cape weed	groundcover	high	—	Farnham, Ayers, and Hasek 1985
Ardisia paniculata	marlberry	shrub	moderate	—	Cal Poly n.d.
Arecastrum romanzoffianum†	queen palm	palm	moderate	—	Skimina 1980
Arecastrum romanzoffianum†	queen palm	palm	low	—	Cal Poly n.d.; Branson and Davis 1965
Artemesia frigida	fringed wormwood	shrub	moderate	—	Donaldson, Hasey, and Davis 1983
Artemesia pycnocephala	sandhill sage	shrub	high	—	Perry 1989
Artemesia stellerana	dusty miller	herbaceous	moderate	—	Glattstein 1989
Arundo donax	giant reed	shrub	high	—	Cal Poly n.d.
Asclepias tuberosa	butterfly weed	herbaceous	moderate	—	Glattstein 1989
Asparagus densiflorus	asparagus fern	herbaceous	moderate	—	Skimina 1980
Asparagus densiflorus 'Sprengeri'	Sprenger asparagus	herbaceous	high	—	Skimina 1980
Atriplex lentiformis ssp. *breweri*	quail bush	groundcover	high	—	Cal Poly n.d.
Atriplex spp.	saltbush	groundcover	high	—	Perry 1989
Baccharis pilularis	prostrate coyote brush	groundcover	high	—	Farnham, Ayers, and Hasek 1985
Baccharis pilularis ssp. *consanguinia*	coyote brush	groundcover	high	high	Cal Poly n.d.
Baccharis sarothroides	desert broom	shrub	high	—	Cal Poly n.d.
Baccharis viminea	mule fat	shrub	high	—	Cal Poly n.d.
Bauhinia purpurea†	orchid tree	tree	moderate	—	Francois 1982
Bauhinia purpurea†	orchid tree	tree	low	low	Cal Poly n.d.; Roewekamp 1941
Berberis × *mentorensis*	mentor barberry	shrub	low	—	Skimina 1980
Berberis thunbergii	Japanese barberry	shrub	low	—	—
Berberis spp.	barberry	shrub	low	—	—
Betula nigra	river birch	tree	low	—	Van Arsdel 1980
Betula papyrifera	paper birch	tree	low	—	—
Betula pendula	European white birch	tree	low	low	Bernstein, Francois, and Clark 1972
Bougainvillea brasiliensis 'Orange King'	orange king bougainvillea	vine	high	—	Skimina 1980
Bougainvillea glabra	paper flower	vine	moderate	—	Cal Poly n.d.
Bougainvillea spectabilis	bougainvillea	vine	high	high	Bernstein, Francois, and Clark 1972; Skimina 1980

TABLE 5.7. Salt and boron tolerance of selected landscape plants, cont.

Scientific name	Common name	Plant type	Salinity tolerance	Boron tolerance	Reference*
Bouteloua gracilis	blue grama	turfgrass	moderate	—	Bernstein 1958; Harivandi 1988
Brahea edulis	Guadalupe palm	palm	moderate	—	Skimina 1980
Brunfelsia pauciflora 'Floribunda'	yesterday-today-tomorrow	shrub	moderate	—	Skimina 1980
Buddleia davidii†	butterfly bush	shrub	low	—	Wu et al. 1999
Butia capitata	pindo palm	palm	high	—	Van Arsdel 1980
Buxus microphylla var. *japonica*†	Japanese boxwood	shrub	high	—	Wu et al. 1999
Buxus microphylla var. *japonica*†	Japanese boxwood	shrub	moderate	high	Bernstein, Francois, and Clark 1972; Skimina 1980
Buxus sempervirens	English boxwood	shrub	low	—	Cal Poly n.d.
Buxus spp.	boxwood	shrub	low	—	—
Calendula officinalis	pot marigold	herbaceous	moderate	moderate	Maas 1984; Morris and Devitt 1990
Callistemon citrinus	lemon bottlebrush	shrub	high	high	Skimina 1980; Francois and Clark 1979
Callistemon rigidus	stiff bottlebrush	shrub	high	—	Perry 1989
Callistemon viminalis†	weeping bottlebush	tree	moderate	—	Bernstein, Francois, and Clark 1972
Callistemon viminalis†	weeping bottlebrush	tree	high	high	Bermstein, Francois, and Clark 1972; Van Arsdel 1980
Callistephus chinensis	China aster	herbaceous	low	moderate	Farnham, Ayers, and Hasek 1985; Maas 1984
Calocedrus decurrens	incense cedar	tree	low	—	Branson and Davis 1965
Calocephalus brownii	cushion bush	herbaceous	high	—	Cal Poly n.d.
Calodendrum capense	Cape chestnut	tree		high	Roewekamp 1941
Camellia japonica	Japanese camellia	shrub	moderate	—	Cal Poly n.d.
Caragana arborescens	Siberian pea tree	tree	high	—	Cornell Coop. Ext. 1999
Carissa macrocarpa	Natal plum	shrub	high	high	Bernstein, Francois, and Clark 1972; Skimina 1980
Carpinus betulus	European hornbeam	tree	low	moderate	Questa 1987
Carpinus spp.	hornbeam	tree	low	—	Butin 1995
Carya illinoensis	pecan	tree	low	low	Van Arsdel 1980
Carya spp.	hickory	tree	low	—	Dirr 1997
Cassia artemisioides	feathery cassia	shrub	high	—	Cal Poly n.d.
Cassia spp.	senna	shrub	high	high	Cal Poly n.d.; San Diego 1966
Casuarina	horsetail tree	tree	moderate	—	Farnham, Ayers, and Hasek 1985; Branson and Davis 1965
Casuarina cunninghamiana	horsetail tree	tree	high	—	Cal Poly n.d.
Casuarina stricta	coast beefwood	tree	high	—	Cal Poly n.d.
Catalpa spp.	catalpa	tree	low	low	Cal Poly n.d.; San Diego 1966
Ceanothus arboreus	feltleaf ceanothus	shrub	high	high	Cal Poly n.d.
Ceanothus gloriosus	Pt. Reyes ceanothus	groundcover	high	high	Cal Poly n.d.
Ceanothus spp.	wild lilac	shrub	—	high	San Diego 1966
Ceanothus thyrsiflorus†	blue blossom	tree	moderate	—	Wu et al. 1999
Ceanothus thyrsiflorus†	blue blossom	tree	high	high	Cal Poly n.d.
Cedrus atlantica	atlas cedar	tree	low	high	Skimina 1980; Questa 1987
Cedrus deodara†	deodar cedar	tree	low	—	Skimina 1980

Scientific name	Common name	Plant type	Salinity tolerance	Boron tolerance	Reference*
Cedrus deodara†	deodar cedar	tree	high	high	Cal Poly n.d.; Questa 1987; Wu et al. 1999
Celosia argenta var. *cristata*	crested cockscomb	herbaceous	low	—	Morris and Devitt 1990; Donaldson, Hasey, and Davis 1983
Celtis australis	European hackberry	tree	low	moderate	Bernstein, Francois, and Clark 1972
Celtis sinensis	Chinese hackberry	tree	low	—	Wu et al. 1999
Cephalanthus occidentalis	buttonwillow	shrub	low	—	Van Arsdel 1980
Ceratonia siliqua	carob tree	tree	moderate	moderate	Roewekamp 1941; Bernstein, Francois, and Clark 1972
Cercidium floridum	blue palo verde	tree	high	—	Cal Poly n.d.
Cercidium spp.	palo verde	tree	high	—	Cal Poly n.d.
Cercis occidentalis	western redbud	tree	high	—	Cal Poly n.d.
Cercis spp.	redbud	tree	low	low	Cal Poly n.d.; San Diego 1966
Cercocarpus betuloides	hardtack	shrub	high	—	Cal Poly n.d.
Cereus peruvianus	apple cactus	herbaceous	moderate	—	Cal Poly n.d.
Chamaerops humilis	European fan palm	palm	high	—	Francois and Clark 1978
Chrysanthemum morifolium 'Bronze Kramer'	mum	herbaceous	moderate	—	Farnham, Ayers, and Hasek 1985
Chrysobalanus icaco	coco plum	tree	moderate	—	Cal Poly n.d.
Cinnamomum camphora†	camphor tree	tree	low	—	Skimina 1980; Branson and Davis 1965
Cinnamomum camphora†	camphor	tree	high	—	Roewekamp 1941
Citrus spp.†	orange, lemon, etc.	tree	low	low	Maas 1984; Eaton 1944
Citrus spp.†	lemon, orange	tree	low	low	San Diego 1966
Clivia miniata 'French Hybrid'	Kaffir lily	shrub	low	—	Skimina 1980
Clytostoma callistegioides	trumpet vine	vine	low	—	Wu et al. 1999
Coccoloba floridana	pigeon plum	tree	high	—	Cal Poly n.d.
Coccolobis uvifera	sea grape	shrub	moderate	—	Cal Poly n.d.
Coccothrinax spp.	palm	palm	high	—	Chase and Broschat 1991
Coccothrinax argentata	broom palm	palm	moderate	—	Cal Poly n.d.
Cocos nucifera	coconut palm	palm	high	—	Chase and Broschat 1991
Codiaeum punctatus	croton	herbaceous	high	—	Cal Poly n.d.
Coleonema spp.	breath of heaven	shrub	high	—	Cal Poly n.d.
Coprosma spp.	mirror plant	shrub	high	—	Farnham, Ayers, and Hasek 1985
Cordyline australis	dracaena palm	palm	low	—	Branson and Davis 1965
Cordyline indivisa	blue dracaena	palm	high	high	Bernstein, Francois, and Clark 1972; Skimina 1980; Francois and Clark 1979
Coreopsis grandiflora	coreopsis	herbaceous	moderate	—	Glattstein 1989
Cornus capitata	evergreen dogwood	shrub	—	low	Roewekamp 1941
Cornus mas	cornelian cherry	shrub	low	—	Wu et al. 1999
Cortaderia selloana	pampas grass	shrub	high	high	Farnham, Ayers, and Hasek 1985
Corylus avellana	European filbert	tree	low	—	Dirr 1997
Cosmos bipinnatus	cosmos	herbaceous	low	—	Morris and Devitt 1990
Cotinus coggygria	smoketree	shrub	moderate	—	Glattstein 1989

TABLE 5.7. Salt and boron tolerance of selected landscape plants, cont.

Scientific name	Common name	Plant type	Salinity tolerance	Boron tolerance	Reference*
Cotoneaster congestus	Pyrenees cotoneaster	shrub	low	—	Francois and Clark 1978
Cotoneaster horizontalis	rock cotoneaster	shrub	low	—	Farnham, Ayers, and Hasek 1985
Cotoneaster lacteus	Parney cotoneaster	shrub	high	—	Cal Poly n.d.
Cotoneaster microphyllus 'Rockspray'	rockspray cotoneaster	groundcover	low	—	Wu et al. 1999
Crassula ovata†	sunset gold	herbaceous	moderate	—	Skimina 1980
Crassula ovata†	jade plant	herbaceous	low	—	Farnham, Ayers, and Hasek 1985
Crataegus × *lavallei*	Carriere hawthorn	tree	low	—	Branson and Davis 1965
Crataegus phaenopyrum	Washington thorn	tree	low	high	Bernstein, Francois, and Clark 1972; Van Arsdel 1980
Crinodendron patagua	lily-of-the-valley tree	tree	—	high	Roewekamp 1941
X *Cupressocyparis leylandii*	Leyland cypress	tree	high	—	Skimina 1980
Cupressus forbesii	Tecate cypress	tree	low	—	Branson and Davis 1965
Cupressus macrocarpa	Monterey cypress	tree	low	—	Branson and Davis 1965
Cupressus sempervirens	Italian cypress	tree	high	high	Cal Poly n.d.; San Diego 1966
Cupressus sempervirens 'Glauca'	blue Italian cypress	tree	moderate	—	Skimina 1980
Cynodon dactylon	bermudagrass	turfgrass	high	—	Bernstein 1958
Cytisus × *praecox* 'Moonlight'	Warminster broom	shrub	low	—	Skimina 1980
Dactylis glomerata	orchard grass	turfgrass	moderate	—	Bernstein 1958
Delosperma alba	white trailing iceplant	groundcover	high	—	Farnham, Ayers, and Hasek 1985; Bernstein, Francois, and Clark 1972
Delphinium spp.	larkspur	herbaceous	low	low	Maas 1984; Eaton 1944
Dianthus barbatus	pinks	herbaceous	low	—	—
Dianthus caryophyllus†	carnation	herbaceous	low	—	Farnham, Ayers, and Hasek 1985
Dianthus caryophyllus†	carnation	herbaceous	high	—	Maas 1984; Morris and Devitt 1990
Dietes iridioides	fortnight lily	shrub	high	—	Skimina 1980
Diospyrus kaki	Japanese persimmon	tree	—	low	Cal Poly n.d.
Diospyros virginiana	persimmon	tree	low	—	Maas 1984; Eaton 1944
Dodonaea viscosa	dodonaea, hop bush	shrub	moderate	—	Francois 1980
Dodonaea viscosa 'Purpurea'	purple hop bush	shrub	moderate	—	Bernstein, Francois, and Clark 1972; Skimina 1980
Dracaena spp.	cordyline	shrub	moderate	—	Cal Poly n.d.
Drosanthemum hispidum	rosea iceplant	groundcover	high	—	Farnham, Ayers, and Hasek 1985; Francois and Clark 1978
Echium fastuosum	pride of Madeira	shrub	high	high	Cal Poly n.d.; San Diego 1966
Elaeagnus angustifolia	Russian olive	shrub	high	—	Dirr 1978; Van Arsdel 1980
Elaeagnus pungens	silverberry	shrub	moderate	low	Bernstein, Francois, and Clark 1972; Francois and Clark 1979
Elaeagnus umbellata	autumn olive	shrub	low	—	Dirr 1978
Ensete ventricosum	Abyssinian banana	herbacenous	low	—	Skimina 1980
Eremochloa ophiuroides	centipedegrass	turfgrass	low	—	Harivandi 1988
Eriobotrya deflexa	bronze loquat	tree	low	—	Branson and Davis 1965

Scientific name	Common name	Plant type	Salinity tolerance	Boron tolerance	Reference*
Eriobotrya japonica	Japanese loquat	tree	low	moderate	Maas 1984
Eriogonum arborescens	Santa Cruz Island buckwheat	shrub	high	—	Cal Poly n.d.
Eriogonum fasciculatum	California buckwheat	shrub	high	—	Cal Poly n.d.
Erythea armata	Brahea blue fan plam	tree	high	—	Cal Poly n.d.
Erythrina caffra	Kaffirboom coral tree	tree	—	low	Roewekamp 1941
Escallonia rubra	escallonia	shrub	modrerate	—	Wu et al. 1999
Eschscholzia californica	California poppy	herbaceous	moderate	moderate	Farnham, Ayers, and Hasek 1985; Glattstein 1989
Eucalyptus camaldulensis	red gum	tree	high	high	Donaldson, Hasey, and Davis 1983; Perry 1989
Eucalyptus citriodora	lemon gum	tree	high	high	Cal Poly n.d.
Eucalyptus cladocalyx	sugar gum	tree	high	high	Cal Poly n.d.
Eucalyptus ficifolia†	red flowering gum	tree	high	—	Cal Poly n.d.
Eucalyptus ficifolia†	red flowering gum	tree	low	moderate	Bernstein, Francois, and Clark 1972; Van Arsdel 1980
Eucalyptus globulus 'Compacta'	dwarf blue gum	tree	high	high	Cal Poly n.d.
Eucalyptus gunnii	cider gum	tree	high	—	Donaldson, Hasey, and Davis 1983
Eucalyptus lehmannii	bushy yate	shrub	high	high	Cal Poly n.d.
Eucalyptus microtheca	flooded box	tree	high	—	Donaldson, Hasey, and Davis 1983
Eucalyptus polyanthemos†	silver dollar gum	tree	high	moderate	Cal Poly n.d.
Eucalyptus polyanthemos†	silver dollar gum	tree		high	Questa 1987
Eucalyptus pulverulenta	silver mountain gum	tree	moderate	—	Cal Poly n.d.
Eucalyptus rudis	swamp gum	tree	high	high	Donaldson, Hasey, and Davis 1983; Perry 1989
Eucalyptus sargentii	salt river mallet	tree	high	—	Donaldson, Hasey, and Davis 1983
Eucalyptus sideroxylon†	red ironbark	tree	high	high	Cal Poly n.d.
Eucalyptus sideroxylon†	red ironbark	tree	moderate	moderate	Bernstein, Francois, and Clark 1972; Van Arsdel 1980
Eucalyptus tereticornis	forest red gum	tree	moderate	—	Cal Poly n.d.
Eucalyptus torquata	coral gum	tree	high	—	Perry 1989
Euonymus alata	winged euonymus	shrub	low	—	—
Euonymus japonica	evergreen euonymus	shrub	high	low	Francois 1980; Francois and Clark 1979
Euonymus japonica 'Grandifolia'	evergreen euonymus	shrub	moderate	low	Skimina 1980
Euonymus japonica 'Grandifolia'	evergreen euonymus	shrub	high	—	Bernstein, Francois, and Clark 1972
Euonymus japonica 'Silver King'	silver king euonymus	shrub	low	—	Skimina 1980
Euphorbia lactea	candelabra cactus	shrub	high	—	Cal Poly n.d.
Euphorbia pulcherrima 'Barbara Ecke'	poinsettia	herbaceous	low	moderate	Maas 1984
Euphorbia tirucalis	milkbush	tree	high	—	Cal Poly n.d.
Euryops pectinatus	golden marguerite	herbaceous	low	—	Wu et al. 1999
Fabiana imbricata	fabiana	shrub	high	—	Cal Poly n.d.
Fagus sylvatica	European beech	tree	low	—	Butin 1995

TABLE 5.7. Salt and boron tolerance of selected landscape plants, cont.

Scientific name	Common name	Plant type	Salinity tolerance	Boron tolerance	Reference*
Fatsia japonica	Japanese aralia	shrub	low	—	Cal Poly n.d.
Feijoa sellowiana	pineapple guava	shrub	low	low	Bernstein, Francois, and Clark 1972; Francois and Clark 1979
Felicia amelloides	felicia	herbaceous	low	—	Farnham, Ayers, and Hasek 1985
Festuca arundinacea	tall fescue	turfgrass	moderate	—	Harivandi 1988
Festuca ovina var. *glauca*	blue fescue	turfgrass	low	—	Farnham, Ayers, and Hasek 1985; Skimina 1980
Festuca rubra	creeping red fescue	turfgrass	low	—	Harivandi 1988
Festuca rubra var. *commutata*	Chewing's fescue	turfgrass	moderate	—	Harivandi 1988
Ficus benjamina	weeping Chinese banyan	tree	low	—	Skimina 1980
Ficus carica	fig	tree	moderate	low	Maas 1984; Eaton 1944
Ficus microcarpa	Indian laurel fig	tree	high	—	Skimina 1980
Ficus rubiginosa	rusty leaf fig	tree	low	—	Branson and Davis 1965
Forsythia × *intermedia*	forsythia	shrub	moderate	—	Wu et al. 1999
Forsythia × *intermedia* 'Spring Glory'	spring glory forsythia	shrub	low	—	Skimina 1980
Fraxinus americana	white ash	tree	low	—	Van Arsdel 1980
Fraxinus oxycarpa 'Raywood'	Raywood ash	tree	low	high	Questa 1987; Wu et al. 1999
Fraxinus pennsylvanica[†] var. *lanceolata*	green ash	tree	high	—	Van Arsdel 1980
Fraxinus pennsylvanica[†] var. *lanceolata*	green ash	tree	moderate	—	—
Fraxinus velutina	Arizona ash	tree	high	high	Van Arsdel 1980; Cal Poly n.d.
Fraxinus velutina var. *glabra* 'Modesto'[†]	Modesto ash	tree		low	Questa 1987
Fraxinus velutina var. *glabra* 'Modesto'[†]	Modesto ash	tree	high	high	Cal Poly n.d.
Fremontia mexicana	southern flannel bush	shrub	high	—	Cal Poly n.d.
Furcraea gigantea	green aloe	shrub	high	—	Cal Poly n.d.
Gardenia augusta	gardenia	shrub	low	moderate	Farnham, Ayers, and Hasek 1985; Maas 1984
Gazania aurantiacum	South African daisy	groundcover	high	—	Farnham, Ayers, and Hasek 1985
Gazania splendens	gazania	groundcover	moderate	—	Cal Poly n.d.
Gazania spp.	gazania	groundcover	high	high	Perry 1989
Gelsemium sempervirens	Carolina jessamine	vine	low	—	Skimina 1980
Genista spp.	broom	shrub	high	—	Cal Poly n.d.
Ginkgo biloba	maidenhair tree, gingko	tree	low	—	Van Arsdel 1980; Donaldson, Hasey, and Davis 1983; Branson and Davis 1965
Gladiolus spp.[†]	gladiolius	herbaceous	low	—	Farnham, Ayers, and Hasek 1985
Gladiolus spp.[†]	gladiolius	herbaceous	moderate	moderate	Maas 1984
Gleditsia triacanthos 'Moraine'	moraine locust	tree	low	—	Branson and Davis 1965
Gleditsia triacanthos f. *inermis*	thornless honey locust	tree	high	low	Questa 1987; Cornell Coop. Ext. 1999

Scientific name	Common name	Plant type	Salinity tolerance	Boron tolerance	Reference*
Grevillea 'Canberra'	Canberra silk oak	tree	high	—	Donaldson, Hasey, and Davis 1983
Grevillea banksii	red silk oak	tree	high	—	Cal Poly n.d.
Grevillea robusta	silk oak	tree	high	—	Cal Poly n.d.
Hakea suaveolens	needle bush	shrub	moderate	—	Farnham, Ayers, and Hasek 1985
Halimodendron halodendron	Russian salt tree	shrub	high	—	Cal Poly n.d.
Hedera canariensis†	Algerian ivy	vine	low	—	Francois 1980; Farnham, Ayers, and Hasek 1985
Hedera canariensis†	Algerian ivy	vine	high	—	Cal Poly n.d.
Hedera canariensis 'Variegata'	variegated Algerian ivy	vine	moderate	—	Cal Poly n.d.
Hedera helix	English ivy	vine	high	—	Cal Poly n.d.
Helianthus annus	sunflower	herbaceous	—	low	Mass 1984
Helianthus debilis Nutt.	cucumber leaf	herbaceous	high	—	Cal Poly n.d.
Heteromeles arbutifolia	toyon	shrub	high	high	Cal Poly n.d.
Hibiscus rosa-sinensis	Chinese hibiscus	shrub	low	high	Francois 1980
Hibiscus rosa-sinensis 'Brillant'	brilliant hibiscus	shrub	low	—	Bernstein, Francois, and Clark 1972; Skimina 1980
Hibiscus rosa-sinensis 'President'	president hibiscus	shrub	high	—	Skimina 1980
Hibiscus rosa-sinensis 'Ross Estey'	Ross Estey hibiscus	shrub	moderate	—	Skimina 1980
Hylocereus undatus	night-blooming cereus	vine	moderate	—	Cal Poly n.d.
Hymenocallis keyensis	spiderlily	herbaceous	moderate	—	Cal Poly n.d.
Hymenosporum flavum	sweetshade	tree	—	high	Roewekamp 1941
Hyophorbe spp.	bottle palm	palm	high	—	Chase and Broschat 1991
Ilex × *altaclarensis* 'Wilsonii'	Wilson holly	shrub	low	—	Skimina 1980
Ilex cornuta†	Chinese holly	shrub	low	—	Francois 1980
Ilex cornuta†	Chinese holly	shrub	high	—	Van Arsdel 1980
Ilex cornuta 'Burfordii'	Burford holly	shrub	low	low	Bernstein, Francois, and Clark 1972; Francois and Clark 1979
Ilex cornuta 'Dazzler'	Chinese holly	shrub	low	—	Skimina 1980
Ilex glabra	inkberry	shrub	moderate	—	Cal Poly n.d.
Ilex opaca	American holly	shrub	low	—	Van Arsdel 1980
Ilex vomitoria†	yaupon	shrub	high	—	Van Arsdel 1980
Ilex vomitoria†	yaupon	shrub	moderate	—	Cal Poly n.d.
Ipomoea pes-caprae	morningglory	groundcover	high	—	Cal Poly n.d.
Isomeris arborea	bladderpod	shrub	high	—	Cal Poly n.d.
Jacaranda acutifolia	jacaranda	tree	high	high	Cal Poly n.d.; Roewekamp 1941
Jacquinia armillaris	barbasco	shrub	moderate	—	Cal Poly n.d.
Jasminum polyanthum	pink jasmine	vine	low	—	Wu et al. 1999
Juglans nigra	black walnut	tree	low	low	Farnham, Ayers, and Hasek 1985
Juglans regia†	English walnut	tree	—	moderate	Questa 1987
Juglans regia†	English walnut	tree	low	low	Maas 1984
Juniperus californica	California juniper	shrub	high	—	Cal Poly n.d.
Juniperus chinensis	spreading juniper	shrub	moderate	—	Bernstein, Francois, and Clark 1972

TABLE 5.7. Salt and boron tolerance of selected landscape plants, cont.

Scientific name	Common name	Plant type	Salinity tolerance	Boron tolerance	Reference*
Juniperus chinensis 'Armstrong'	Armstrong juniper	shrub	moderate	low	Branson and Davis 1965
Juniperus chinensis 'Kazuka'	Hollywood juniper	shrub	moderate	high	Questa 1987
Juniperus chinensis 'Pfitzerana'	Pfitzer juniper	shrub	moderate	high	Skimina 1980
Juniperus chinensis 'Robusta Green'	Chinese juniper	shrub	high	—	Skimina 1980
Juniperus communis 'Hibernica'	Irish juniper	shrub	moderate	low	Cal Poly n.d.
Juniperus conferta	shore juniper	groundcover	moderate	—	Branson and Davis 1965
Juniperus horizontalis	creeping juniper	groundcover	moderate	—	Glattstein 1989
Juniperus procumbens	Japanese garden juniper	groundcover	high	—	Cal Poly n.d.
Juniperus sabina 'Tamariscifolia'	tam juniper	groundcover	high	—	Cal Poly n.d.
Juniperus scopulorum 'Moffettii'	Moffett's juniper	shrub	low	—	Farnham, Ayers, and Hasek 1985
Juniperus virginiana	eastern red cedar	shrub	moderate	—	Dirr 1997
Juniperus virginiana 'Skyrocket'	skyrocket juniper	shrub	high	—	Wu et al. 1999
Kalanchoe spp.	kalanchoe	herbaceous	moderate	—	Cal Poly n.d.
Koelreuteria paniculata†	goldenrain tree	tree	high	—	Roewekamp 1941; Cornell Coop. Ext. 1999
Koelreuteria paniculata†	goldenrain tree	tree	low	—	Wu et al. 1999
Kopsia arborea	blume	tree	moderate	—	Cal Poly n.d.
Lagerstroemia indica	crape myrtle	tree	low	moderate	Francois 1982
Lagunaria patersonii	primrose tree	tree	high	high	Cal Poly n.d.; Roewekamp 1941
Lampranthus productus	purple iceplant	groundcover	high	—	Francois and Clark 1978; Francois 1980
Lampranthus spectabilis	trailing iceplant	groundcover	high	—	Farnham, Ayers, and Hasek 1985
Lantana sp. 'Confetti'	confetti lantana	shrub	low	—	Skimina 1980
Lantana camara	yellow lantana	shrub	moderate	low	Bernstein, Francois, and Clark 1972; Francois and Clark 1979
Lantana montevidensis	trailing lantana	groundcover	moderate	high	Cal Poly n.d.; San Diego 1966
Lathyrus japonicus	sweet pea	herbaceous	high	—	Cal Poly n.d.
Lathyrus odoratus	sweet pea	herbaceous	high	high	Maas 1984; Eaton 1944
Laurus nobilis	Grecian laurel	tree	—	high	Questa 1987
Lavatera assurgentiflora	tree mallow	shrub	high	—	Perry 1989
Leptospermum laevigatum†	Australian tea tree	shrub	low	—	Farnham, Ayers, and Hasek 1985
Leptospermum laevigatum†	Australian tea tree	shrub	high	—	Cal Poly n.d.
Leucophyllum frutescens†	Texas sage	shrub	moderate	moderate	Maas 1984
Leucophyllum frutescens†	Texas sage	shrub	high	high	Francois and Clark 1978; Francois 1980
Ligustrum spp.	privet	shrub	high	—	Van Arsdel 1980

Scientific name	Common name	Plant type	Salinity tolerance	Boron tolerance	Reference*
Ligustrum japonicum[†]	wax-leaf privet	shrub	—	high	Roewekamp 1941
Ligustrum japonicum[†]	wax-leaf privet	shrub	moderate	low	Skimina 1980; Francois and Clark 1979; Branson and Davis 1965
Ligustrum lucidum[†]	glossy privet	shrub	moderate	moderate	Bernstein, Francois, and Clark 1972; Branson and Davis 1965
Ligustrum lucidum[†]	glossy privet	shrub	—	high	Questa 1987
Limonium perezii	sea lavender	herbaceous	low	—	Farnham, Ayers, and Hasek 1985
Liquidambar styraciflua[†]	sweetgum	tree	low	—	Van Arsdel 1980; Wu et al. 1999
Liquidambar styraciflua[†]	sweetgum	tree	moderate	moderate	Francois 1982
Liquidambar styraciflua 'Festival'	festival sweetgum	tree	—	low	Questa 1987
Liriodendron tulipifera	tulip tree	tree	low	moderate	Francois 1982
Livistona chinensis	Chinese fan palm	palm	moderate	—	Cal Poly n.d.
Lobularia maritima	sweet alyssum	herbaceous	moderate	—	Morris and Devitt 1990
Lolium perenne	perennial ryegrass	turfgrass	moderate	—	Bernstein 1958; Harivandi 1988
Lonicera hildebrandiana	giant honeysuckle	vine	high	—	Cal Poly n.d.
Lonicera japonica[†]	honeysuckle	vine	high	—	Cal Poly n.d.
Lonicera japonica[†]	honeysuckle	vine	moderate	—	—
Lonicera xylosteum	European fly honeysuckle	vine	high	—	—
Lotus corniculatus	bird's foot trefoil	groundcover	high	—	Cal Poly n.d.; Bernstein 1958
Lupinus hartwegii	lupine	herbaceous	—	low	Eaton 1944
Maclura pomifera	Osage orange	tree	moderate	—	Van Arsdel 1980
Magnolia grandiflora[†]	southern magnolia	tree	moderate	—	Francois 1982; Van Arsdel 1980
Magnolia grandiflora[†]	southern magnolia	tree	low	low	Skimina 1980; Roewekamp 1941
Magnolia spp. (deciduous)	magnolia	tree	low	low	San Diego 1966
Mahonia aquifolium	Oregon grape	shrub	low	low	Francois and Clark 1978; Francois 1980
Mahonia aquifolium 'Compacta'	compact Oregon grape	shrub	low	—	Skimina 1980
Mahonia nevinii	Nevin mahonia	shrub	high	—	Cal Poly n.d.
Mahonia pinnata	California holly grape	shrub	low	—	Wu et al. 1999
Malephora crocea	croceum iceplant	groundcover	high	—	—
Malus spp.	crabapple	tree	—	high	Questa 1987
Malus sylvestris	apple	tree	low	—	Maas 1984
Matthiola incana	stock	herbaceous	moderate	—	Farnham, Ayers, and Hasek 1985
Maytenus boaria	mayten	tree	—	high	Questa 1987
Melaleuca armillaris	drooping melaleuca	tree	high	—	Cal Poly n.d.
Melaleuca nesophila	pink melaleuca	tree	high	—	Cal Poly n.d.; Perry 1989
Melaleuca quinquenervia	cajeput tree	tree	moderate	—	Farnham, Ayers, and Hasek 1985
Mesembryanthemum spp. and var.	ice plant	groundcover	high	—	Cal Poly n.d.
Metrosideros excelsus	New Zealand christmas tree	tree	high	—	Perry 1989; Branson and Davis 1965
Monstera deliciosa	philodendron	herbaceous	low	—	Cal Poly n.d.
Morus alba	white mulberry	tree	—	moderate	Questa 1987

TABLE 5.7. Salt and boron tolerance of selected landscape plants, cont.

Scientific name	Common name	Plant type	Salinity tolerance	Boron tolerance	Reference*
Morus alba 'Fruitless'	fruitless white mulberry	tree	moderate	low	Cal Poly n.d.; San Diego 1966
Musa acuminata	banana	shrub	low	—	Cal Poly n.d.
Myoporum laetum†	myoporum	shrub	high	—	Farnham, Ayers, and Hasek 1985; Perry 1989
Myoporum laetum†	myoporum	shrub	moderate	—	Cal Poly n.d.; Branson and Davis 1965
Myoporum parvifolium	myoporum	groundcover	high	—	Farnham, Ayers, and Hasek 1985; Donaldson, Hasey, and Davis 1983
Myrica pensylvanica	bayberry	shrub	moderate	—	Glattstein 1989
Myrtus communis†	myrtle	shrub	high	moderate	Cal Poly n.d.; San Diego 1966
Myrtus communis†	myrtle	shrub	moderate	—	Wu et al. 1999
Myrtus communis 'Compacta'	dwarf myrtle	shrub	high	—	Cal Poly n.d.
Nandina domestica	heavenly bamboo	shrub	low	—	Bernstein, Francois, and Clark 1972; Skimina 1980
Nerium oleander	oleander	shrub	high	high	Bernstein, Francois, and Clark 1972; Francois and Clark 1979
Nerium oleander 'Cherry Ripe'	cherry ripe oleander	shrub	moderate	—	Skimina 1980
Ochrosia elliptica	labill	shrub	moderate	—	Cal Poly n.d.
Oenothera speciosa childsii	Mexican evening primrose	herbaceous	high	—	Cal Poly n.d.
Olea africana	African olive	shrub	moderate	—	Donaldson, Hasey, and Davis 1983
Olea europaea†	olive	tree	high	high	Bernstein, Francois, and Clark 1972
Olea europaea†	olive	tree	moderate	moderate	Maas 1984; Farnham, Ayers, and Hasek 1985
Olea europaea 'Montra'	dwarf olive	shrub	high	—	Wu et al. 1999
Ophiopogon jaburan	giant lilyturf	herbaceous	moderate	—	Skimina 1980
Ophiopogon japonicus	mondo grass	herbaceous	low	—	Skimina 1980
Ostrya virginiana	eastern hop hornbeam	tree	low	—	Cornell Coop. Ext. 1999
Oxalis bowiei	oxalis	herbaceous	—	high	Eaton 1944
Pachysandra terminalis	Japanese spurge	groundcover	low	—	Farnham, Ayers, and Hasek 1985
Parkinsonia aculeata	Mexican palo verde	tree	high	high	Cal Poly n.d.; San Diego 1966
Parthenocissus quinquefolia	Virginia creeper	vine	moderate	—	Cal Poly n.d.
Paspalum vaginatum	seashore paspalum	turfgrass	high	—	Harivandi 1988
Passiflora edulis	passion fruit	vine	low	—	Maas 1984
Pelargonium × *hortorum*	geranium	herbaceous	low	low	Farnham, Ayers, and Hasek 1985; Maas 1984
Pelargonium domesticum 'Martha Washington'	Martha Washington geranium	herbaceous	low	low	San Diego 1966
Pelargonium peltatum	ivy geranium	herbaceous	moderate	moderate	San Diego 1966
Persea americana†	avocado	tree	high	—	Maas 1984
Persea americana†	avocado	tree	low	low	Maas 1984

Scientific name	Common name	Plant type	Salinity tolerance	Boron tolerance	Reference*
Petunia hybrida	petunia	herbaceous	high	—	Morris and Devitt 1990
Philodendron bipinnatifidum	tree philodendron	shrub	low	—	Cal Poly n.d.
Philodendron selloum	split-leaf philodendron	shrub	moderate	—	Skimina 1980
Phoenix canariensis	Canary Island palm	palm	high	high	Farnham, Ayers, and Hasek 1985
Phoenix dactylifera	date palm	palm	high	high	Maas 1984; Farnham, Ayers, and Hasek 1985
Phoenix loureiri	pigmy date palm	palm	low	—	Cal Poly n.d.
Phormium tenax†	New Zealand flax	shrub	moderate	—	Farnham, Ayers, and Hasek 1985
Phormium tenax†	New Zealand flax	shrub	high	—	Cal Poly n.d.
Phormium tenax 'Atropurpureum'	purple New Zealand flax	shrub	low	—	Skimina 1980
Photinia × *fraseri*	Fraser's photinia	shrub	low	low	Farnham, Ayers, and Hasek 1985; Francois and Clark 1979
Photinia serrulata	Chinese photinia	shrub	low	low	Cal Poly n.d.; San Diego 1966
Phyla nodiflora	lippia	groundcover	high	—	Farnham, Ayers, and Hasek 1985
Picea abies	Norway spruce	tree	low	—	Dirr 1997
Picea glauca	white spruce	tree	low	—	Dirr 1997
Picea pungens	blue spruce	tree	high	—	Dirr 1997
Pinus brutia	brutia pine	tree	high	—	Donaldson, Hasey, and Davis 1983
Pinus canariensis	Canary Island pine	tree	—	high	Questa 1987
Pinus cembroides	Mexican piñon pine	tree	high	—	Wu et al. 1999
Pinus eldarica	Afghan pine	tree	moderate	—	Donaldson, Hasey, and Davis 1983
Pinus elliottii	slash pine	tree	moderate	—	Van Arsdel 1980
Pinus halepensis	Aleppo pine	tree	high	moderate	Francois and Clark 1979; Perry 1989
Pinus nigra	Austrian black pine	tree	high	—	Van Arsdel 1980
Pinus pinea	Italian stone pine	tree	high	moderate	Francois 1982
Pinus ponderosa	ponderosa pine	tree	moderate	—	Dirr 1997
Pinus radiata	Monterey pine	tree	high	high	Cal Poly n.d.; Questa 1987
Pinus resinosa	Norway pine	tree	low	—	Dirr 1997
Pinus strobus	eastern white pine	tree	low	—	Cornell Coop. Ext. 1999
Pinus sylvestris	Scotch pine	tree	moderate	—	Van Arsdel 1980
Pinus taeda	loblolly pine	tree	moderate	—	Van Arsdel 1980
Pinus thunbergiana	Japanese black pine	tree	moderate	moderate	Francois and Clark 1978; Skimina 1980
Pistacia chinensis†	Chinese pistache	tree	moderate	low	Bernstein, Francois, and Clark 1972; Van Arsdel 1980
Pistacia chinensis†	Chinese pistache	tree	low	—	Wu et al. 1999
Pittosporum crassifolium	evergreen pittosporum	shrub	high	—	Farnham, Ayers, and Hasek 1985; Perry 1989
Pittosporum phillyraeoides†	willow pittosporum	tree	moderate	—	Farnham, Ayers, and Hasek 1985
Pittosporum phillyraeoides†	willow pittosporum	tree	high	—	Perry 1989
Pittosporum tobira†	mock orange	shrub	low	high	Francois 1980

TABLE 5.7. Salt and boron tolerance of selected landscape plants, cont.

Scientific name	Common name	Plant type	Salinity tolerance	Boron tolerance	Reference*
Pittosporum tobira†	mock orange	shrub	moderate	low	Bernstein, Francois, and Clark 1972; Francois and Clark 1979; Branson and Davis 1965
Pittosporum tobira†	mock orange	shrub	high	—	Wu et al. 1999
Pittosporum tobira 'Variegata'	variegated mock orange	shrub	low	—	Skimina 1980
Pittosporum undulatum†	Victorian box	shrub	—	low	Roewekamp 1941
Pittosporum undulatum†	Victorian box	shrub	—	high	Questa 1987
Platanus × acerifolia	London plane	tree	high	moderate	Cornell Coop. Ext. 1999; Questa 1987
Platanus occidentalis	American sycamore	tree	low	—	Van Arsdel 1980
Platanus racemosa	California sycamore	tree	high	high	Cal Poly n.d.; Questa 1987
Platycladus orientalis†	oriental arborvitae	tree	low	—	Maas 1984
Platycladus orientalis†	oriental arborvitae	tree	moderate	moderate	Bernstein, Francois, and Clark 1972
Platycladus orientalis 'Aureus Nana'	dwarf golden arborvitae	shrub	—	—	Skimina 1980
Plumbago auriculata	Cape plumbago	shrub	high	—	Wu et al. 1999
Plumeria rubra	frangipani plumeria	tree	moderate	—	Cal Poly n.d.
Poa annua	annual bluegrass	turfgrass	low	—	Harivandi 1988
Poa pratensis	Kentucky bluegrass	turfgrass	low	—	Harivandi 1988
Poa trivialis	rough bluegrass	turfgrass	low	—	Harivandi 1988
Podocarpus gracilior	yew pine	tree	—	high	Questa 1987
Podocarpus macrophyllus†	yew pine	tree	low	moderate	Maas 1984; Francois 1980
Podocarpus macrophyllus†	yew pine	tree	high	—	Van Arsdel 1980
Podocarpus macrophyllus var. *maki*	shrubby Japanese yew pine	shrub	low	moderate	Francois and Clark 1978, 1979; Skimina 1980
Populus × canadensis	Carolina poplar	tree	high	—	Cal Poly n.d.
Populus alba	white poplar	tree	high	—	—
Populus deltoides	eastern cottonwood	tree	moderate	—	—
Populus fremontii†	western cottonwood	tree	moderate	—	Van Arsdel 1980
Populus fremontii†	western cottonwood	tree	high	—	Cal Poly n.d.
Populus nigra 'Italica'	Lombardy poplar	tree	low	—	—
Populus spp.	cottonwood	tree	—	high	Questa 1987
Portulaca grandiflora	rose moss	groundcover	high	—	Morris and Devitt 1990
Prosopis grandulosa var. *torreyana*	honey mesquite	tree	high	—	Cal Poly n.d.
Prunus × blireiana	flowering plum	tree	moderate	—	Branson and Davis 1965
Prunus armeniaca	apricot	tree	low	—	Maas 1984
Prunus avium	cherry	tree	low	low	Maas 1984; Eaton 1944
Prunus besseyi	sand cherry	tree	low	—	Dirr 1978
Prunus caroliniana	Carolina laurel cherry	shrub	moderate	high	Roewekamp 1941; Bernstein, Francois, and Clark 1972
Prunus cerasifera	cherry plum	tree	moderate	—	Francois 1982

Scientific name	Common name	Plant type	Salinity tolerance	Boron tolerance	Reference*
Prunus cerasifera 'Atropurpurea'	purple-leaf plum	tree	—	high	Questa 1987
Prunus domestica	plum, prune	tree	low	low	Maas 1984
Prunus dulcis	almond	tree	low	—	Maas 1984
Prunus ilicifolia	hollyleaf cherry	tree	high	high	Cal Poly n.d.; San Diego 1966
Prunus lyonii	Catalina cherry	tree	high	high	Bernstein, Francois, and Clark 1972
Prunus persica	peach	tree	low	low	Maas 1984; Eaton 1944
Prunus sargentii	Sargent cherry	tree	high	—	Cornell Coop. Ext. 1999
Prunus serrulata	flowering cherry	tree	—	moderate	Questa 1987
Prunus tomentosa	Nanking cherry	tree	low	—	Dirr 1978
Pseudotsuga menziesii	Douglas fir	tree	moderate	—	Dirr 1997
Psidium cattleianum	strawberry guava	shrub	high	high	Cal Poly n.d.; San Diego 1966
Puccinellia spp.	alkaligrass	turfgrass	high	—	Harivandi 1988
Punica granatum[†]	dwarf pomegranate	shrub	low	—	Farnham, Ayers, and Hasek 1985
Punica granatum[†]	pomegranate	tree	moderate	—	Maas 1984
Punica granatum 'Wonderful'	wonderful pomegranate	shrub	high	high	Cal Poly n.d.
Pyracantha cernato-serrata	pyracantha	shrub	moderate	—	Francois 1980
Pyracantha cernato-serrata 'Graberi'[†]	pyracantha	shrub	high	—	Bernstein, Francois, and Clark 1972
Pyracantha cernato-serrata 'Graberi'	pyracantha	shrub	moderate	—	Cal Poly n.d.
Pyracantha coccinea[†]	firethorn	shrub	high	—	Van Arsdel 1980
Pyracantha coccinea[†]	firethorn	shrub	moderate	—	—
Pyracantha koidzumii 'Victory'	victory pyracantha	shrub	low	—	Skimina 1980
Pyrus calleryana	callery pear	tree	high	—	Cornell Coop. Ext. 1999
Pyrus communis	pear	tree	low	—	Maas 1984
Pyrus kawakamii	evergreen pear	tree	high	high	Bernstein, Francois, and Clark 1972
Quercus agrifolia	coast live oak	tree	moderate	high	Questa 1987; Wu et al. 1999
Quercus bicolor	swamp white oak	tree	low	—	Cornell Coop. Ext. 1999
Quercus ilex	holly oak	tree	—	moderate	Questa 1987
Quercus lobata	valley oak	tree	—	high	Questa 1987
Quercus marilandica	blackjack oak	tree	moderate	—	Van Arsdel 1980
Quercus palustris[†]	pin oak	tree	moderate	—	Van Arsdel 1980
Quercus palustris[†]	pin oak	tree	low	—	Cornell Coop. Ext. 1999
Quercus robur	English oak	tree	high	—	Cornell Coop. Ext. 1999
Quercus rubra	northern red oak	tree	high	—	Cornell Coop. Ext. 1999
Quercus suber	cork oak	tree	—	high	Questa 1987
Quercus virginiana	southern live oak	tree	high	—	Van Arsdel 1980
Rhamnus alaternus	Italian buckthorn	shrub	low	—	Farnham, Ayers, and Hasek 1985
Rhamnus californica	coffeeberry	shrub	high	—	Cal Poly n.d.
Rhaphiolepis indica[†]	India hawthorn	shrub	moderate	—	Farnham, Ayers, and Hasek 1985; Francois and Clark 1978, 1989
Rhaphiolepis indica[†]	India hawthorn	shrub	high	—	Wu et al. 1999

TABLE 5.7. Salt and boron tolerance of selected landscape plants, cont.

Scientific name	Common name	Plant type	Salinity tolerance	Boron tolerance	Reference*
Rhaphiolepis indica 'Enchantress'	enchantress India hawthorn	shrub	moderate	high	Skimina 1980; Francois and Clark 1979
Rhaphiolepis umbellata	yeddo hawthorn	shrub	moderate	—	Cal Poly n.d.
Rhododendron spp.	rhododendron	shrub	low	—	Cal Poly n.d.
Rhododendron spp.	azalea	shrub	low	—	Farnham, Ayers, and Hasek 1985
Rhus integrifolia	lemonade berry	shrub	high	—	Cal Poly n.d.
Rhus ovata	sugar bush	shrub	high	—	Cal Poly n.d.
Rhus trilobata	squaw bush	shrub	high	—	—
Ribes spp.	currant	shrub	low	—	Maas 1984
Robinia pseudoacacia	black locust	tree	high	—	Cornell Coop. Ext. 1999
Rosa spp.	rose	shrub	low	—	Van Arsdel 1980; Francois 1980
Rosa chinensis 'Gloire des Rosomanes'	ragged robin rose	shrub	—	moderate	Farnham, Ayers, and Hasek 1985
Rosa hybrida	hybrid rose	shrub	high	—	Cal Poly n.d.
Rosa multiflora	Japanese rose	shrub	low	—	—
Rosa rugosa	rugosa rose	shrub	high	—	Dirr 1978
Rosa 'Grenoble'	rose	shrub	low	—	Bernstein, Francois, and Clark 1972
Rosmarinus officinalis[†]	rosemary	groundcover	low	moderate	Maas 1984; Francois and Clark 1979
Rosmarinus officinalis[†]	rosemary	groundcover	—	low	Maas 1984
Rosmarinus officinalis[†]	rosemary	groundcover	high	high	Cal Poly n.d.
Rosmarinus officinalis 'Lockwood de Forest'	creeping rosemary	groundcover	high	—	Bernstein, Francois, and Clark 1972
Rosmarinus officinalis 'Prostratus'	prostrate rosemary	groundcover	moderate	—	Cal Poly n.d.
Sabal palmetto	cabbage palm	palm	moderate	—	Cal Poly n.d.
Salix alba 'Vitellina'	golden willow	tree	moderate	—	—
Salix babylonica	weeping willow	tree	high	high	Cal Poly n.d.; Questa 1987
Salix matsudana 'Torutosa'	corkscrew willow	tree	—	high	Questa 1987
Salix purpurea var. *nana*	dwarf blue willow	shrub	low	—	—
Salvia spp.	salvia, sage	herbaceous	high	—	Cal Poly n.d.
Sambucus nigra	elderberry	shrub	low	—	Wu et al. 1999
Sansevieria spp.	snake plant	herbaceous	moderate	—	Cal Poly n.d.
Sapium sebiferum	Chinese tallow tree	tree	high	high	Van Arsdel 1980; Questa 1987
Schinus molle	California pepper tree	tree	high	high	Bernstein, Francois, and Clark 1972
Schinus terebinthifolius	Brazilian pepper tree	tree	high	high	Perry 1989; Questa 1987
Sedum spp.	sedum	herbaceous	moderate	—	Cal Poly n.d.
Sequoia sempervirens	coast redwood	tree	—	low	Questa 1987
Serenoa repens	saw palmetto	palm	high	—	Chase and Broschat 1991
Shepherdia argentea	buffalo berry	shrub	moderate	—	—
Simmondsia chinensis	jojoba	shrub	high	—	Maas 1984
Sophora japonica	Japanese pagoda tree	tree	high	moderate	Butin 1995; Questa 1987
Sorbus aucuparia	European mountain ash	tree	low	—	Butin 1995

Scientific name	Common name	Plant type	Salinity tolerance	Boron tolerance	Reference*
Spartium junceum	Spanish broom	shrub	high	—	Skimina 1980
Spiraea vanhouttei	spirea	shrub	low	—	—
Stachys betonicifolia	lamb's ear	herbaceous	moderate	—	Glattstein 1989
Stapelia gigantea	starfish flower	herbaceous	moderate	—	Cal Poly n.d.
Stenotaphrum secundatum	St. Augustinegrass	turfgrass	high	—	Harivandi 1988
Strelitzia reginae†	bird of paradise	shrub	low	—	Farnham, Ayers, and Hasek 1985
Strelitzia reginae†	bird of paradise	shrub	moderate	moderate	Cal Poly n.d.; San Diego 1966
Syzygium paniculatum†	brush cherry	shrub	high	—	Francois and Clark 1978
Syzygium paniculatum†	brush cherry	shrub	moderate	moderate	Maas 1984; Skimina 1980; Francois and Clark 1979
Tamarix africanus	tamarisk	shrub	moderate	—	Cal Poly n.d.
Tamarix aphylla	athel tree	shrub	high	high	Farnham, Ayers, and Hasek 1985
Tamarix gallica	tamarisk	shrub	high	—	—
Tamarix parviflora	pink tamarisk	shrub	high	high	Cal Poly n.d.; San Diego 1966
Taxodium distichum	bald cypress	tree	high	—	Branson and Davis 1965
Tetrastigma harmandi	Harmand's grape	vine	moderate	—	Cal Poly n.d.
Teucrium chamadrys	germander	shrub	moderate	moderate	Cal Poly n.d.; San Diego 1966
Thrinax microcarpa	key palm	palm	high	—	Chase and Broschat 1991
Thrinax parviflora	mountain thatch palm	palm	high	—	Chase and Broschat 1991
Thuja occidentalis	American arborvitae	shrub	moderate	—	Dirr 1997
Thuja orientalis	oriental arborvitae	shrub	moderate	—	Cal Poly n.d.
Thuja orientalis 'Aurea'	dwarf golden arborvitae	shrub	high	—	Branson and Davis 1965
Thunbergia erecta	bush clock vine	vine	low	—	Cal Poly n.d.
Tilia × *euchlora*	Crimean linden	tree	low	—	Cornell Coop. Ext. 1999
Tilia americana 'Redmond'	American basswood	tree	low	—	Cornell Coop. Ext. 1999
Tilia cordata	little-leaf linden	tree	low	—	Cornell Coop. Ext. 1999
Tilia spp.	linden	tree	low	—	Van Arsdel 1980
Trachelospermum jasminoides†	star jasmine	groundcover	low	—	Bernstein, Francois, and Clark 1972; Francois 1980
Trachelospermum jasminoides†	star jasmine	groundcover	moderate	—	Cal Poly n.d.
Trachycarpus fortunei	windmill palm	palm	moderate	—	Van Arsdel 1980
Tribulus terrestris	puncturevine	vine	high	—	Cal Poly n.d.
Trifolium fragiferum	strawberry clover	groundcover	moderate	—	Bernstein 1958
Trifolium repens	white Dutch clover	groundcover	low	—	Bernstein 1958
Tropaeolum majus	nasturtium	herbaceous	moderate	—	Glattstein 1989
Tsuga canadensis	eastern hemlock	tree	low	—	Dirr 1997
Ulmus alata	winged elm	tree	low	—	Van Arsdel 1980
Ulmus americana	American elm	tree	low	low	Van Arsdel 1980, Maas 1984
Ulmus carpinifolia	smoothleaf elm	tree	high	—	—
Ulmus crassifolia	cedar elm	tree	low	—	Van Arsdel 1980
Ulmus glabra	Scotch elm, wych elm	tree	high	—	Butin 1995
Ulmus parvifolia†	Chinese evergreen elm	tree	—	low	Roewekamp 1941
Ulmus parvifolia†	Chinese evergreen elm	tree	—	moderate	Questa 1987

TABLE 5.7. Salt and boron tolerance of selected landscape plants, cont.

Scientific name	Common name	Plant type	Salinity tolerance	Boron tolerance	Reference*
Ulmus pumila†	Siberian elm	tree	high	moderate	Van Arsdel 1980; Roewekamp 1941
Ulmus pumila†	Siberian elm	tree	—	low	Questa 1987
Uniola paniculata	sea oats	turfgrass	high	—	Cal Poly n.d.
Viburnum tinus	laurustinus	shrub	low	low	Francois 1980
Viburnum tinus 'Robustum'	laurustinus	shrub	moderate	—	Bernstein, Francois, and Clark 1972; Branson and Davis 1965
Vinca major	periwinkle	groundcover	high	—	Cal Poly n.d.
Vinca minor	dwarf periwinkle	groundcover	low	—	Farnham, Ayers, and Hasek 1985
Viola ordorata	violet	herbaceous	—	low	Eaton 1944
Viola tricolor	pansy	herbaceous	low	low	Maas 1984; Eaton 1944
Vitex lucens	New Zealand chaste tree	tree	—	high	Roewekamp 1941
Washingtonia filifera†	California fan palm	palm	moderate	—	Van Arsdel 1980; Branson and Davis 1965
Washingtonia filifera†	California fan palm	palm	high	—	Wu et al. 1999
Washingtonia robusta†	Mexican fan palm	palm	low	—	Skimina 1980
Washingtonia robusta†	Mexican fan palm	palm	moderate	—	Branson and Davis 1965
Woodwardia spp.	chain fern	shrub	moderate	—	Cal Poly n.d.
Xylosma congestum	shiny xylosma	shrub	moderate	low	Bernstein, Francois, and Clark 1972; Branson and Davis 1965
Yucca aloifolia	Spanish bayonet	shrub	high	—	Skimina 1980
Yucca filamentosa†	Adam's needle	shrub	low	—	Skimina 1980
Yucca filamentosa†	Adam's needle	shrub	high	—	Cal Poly n.d.; Glattstein 1989
Yucca spp.	yucca	shrub	high	—	Van Arsdel 1980
Zamia spp.	pumila	palm	moderate	—	Cal Poly n.d.
Zanthoxylum fagara	wild lime	shrub	moderate	—	Cal Poly n.d.
Zelkova serrata	sawleaf zelkova	tree	—	moderate	Questa 1987
Zinnia elegans	zinnia	herbaceous	low	low	Maas 1984; Eaton 1944
Ziziphus jujuba	jujube	tree	high	—	Perry 1989
Zoysia tenuifolia	zoysiagrass	turfgrass	high	—	Cal Poly n.d.

Source: After Matheny and Clark 1998.
Note:
— = not available.
*Refer to table 5.10 for methods and criteria used in evaluating salinity and boron tolerance.
†Variable reports in tolerance rating for species.

TABLE 5.8. Salt tolerance of selected landscape plants (see table 5.7 for common names and references; bold-face type indicates plants listed in more than one category)

High	Moderate	Low
	Groundcovers	
Acacia redolens	*Gazania splendens*	*Cotoneaster microphyllus* ‘Rockspray’
Arctotheca calendula	*Juniperus conferta*	*Ophiopogon japonicus*
Atriplex lentiformis ssp. *breweri*	*Juniperus horizontalis*	*Pachysandra terminalis*
Atriplex spp.	*Lantana montevidensis*	***Rosmarinus officinalis***
Baccharis pilularis	*Ophiopogon jaburan*	***Trachelospermum jasminoides***
Baccharis pilularis ssp. *consanguinia*	*Rosmarinus officinalis* ‘Prostratus’	*Trifolium repens*
Ceanothus gloriosus	***Trachelospermum jasminoides***	*Vinca minor*
Delosperma alba	*Trifolium fragiferum*	
Drosanthemum hispidum		
Gazania aurantiacum		
Gazania spp.		
Ipomoea pes-caprae		
Juniperus procumbens		
Juniperus sabina ‘Tamariscifolia’		
Lampranthus productus		
Lampranthus spectabilis		
Lotus corniculatus		
Malephora crocea		
Mesembryanthemum spp. and var.		
Myoporum parvifolium		
Phyla nodiflora		
Portulaca grandiflora		
Rosmarinus officinalis		
Rosmarinus officinalis ‘Lockwood de Forest’		
Vinca major		
	Herbaceous Plants	
Aloe vera	*Achillea* spp.	*Alyssum saxatile*
Asparagus densiflorus Sprengeri’	*Agapanthus orientalis*	*Antirrhinum majus*
Calocephalus brownii	*Agapanthus umbellatus*	*Callistephus chinensis*
Codiaeum punctatus	*Ageratum* spp.	*Celosia argentea* var. *cristata*
Dianthus caryophyllus	*Artemesia stellerana*	*Cosmos bipinnatus*
Helianthus debilis Nutt.	*Asclepias tuberosa*	***Crassula ovata***
Lathyrus japonicus	*Asparagus densiflorus*	*Delphinium* spp.
Lathyrus odoratus	*Calendula officinalis*	*Dianthus barbatus*
Oenothera speciosa childsii	*Cereus peruvianus*	***Dianthus caryophyllus***
Petunia hybrida	*Chrysanthemum morifolium* ‘Bronze Kramer’	*Ensete ventricosum*
Salvia spp.	*Coreopsis grandiflora*	*Euphorbia pulcherrima* ‘Barbara Ecke’
	Crassula ovata	*Euryops pectinatus*
	Eschscholzia californica	*Felicia amelloides*
	***Gladiolus* spp.**	***Gladiolus* spp.**
	Hymenocallis keyensis	*Limonium perezii*

TABLE 5.8. Salt tolerance of selected landscape plants, cont.

High	Moderate	Low
	Herbaceous Plants, cont.	
	Kalanchoe spp.	*Monstera deliciosa*
	Lobularia maritima	*Pelargonium* × *hortorum*
	Matthiola incana	*Pelargonium domesticum* 'Martha Washington'
	Pelargonium peltatum	*Viola tricolor*
	Sansevieria spp.	
	Sedum spp.	
	Stachys betonicifolia	
	Stapelia gigantea	
	Tropaeolum majus	
	Zamia spp.	
	Vines	
Bougainvillea spectabilis	*Allamanda cathartica*	*Clytostoma callistegioides*
Bougainvillea brasiliensis 'Orange King'	*Bougainvillea glabra*	*Gelsemium sempervirens*
Hedera canariensis	*Hedera canariensis* 'Variegata'	***Hedera canariensis***
Hedera helix	***Lonicera japonica***	*Jasminum polyanthum*
Lonicera hildebrandiana	*Parthenocissus quinquefolia*	*Passiflora edulis*
Lonicera japonica	*Tetrastigma harmandi*	*Thunbergia erecta*
Lonicera xylosteum		
Tribulus terrestris		
	Shrubs	
Agave spp.	*Agave attenuta*	*Abelia* × *grandiflora*
Aloe arborescens	*Ardisia paniculata*	*Abelia* × *grandiflora* 'Edward Goucher'
Artemesia pycnocephala	*Artemesia frigida*	*Acacia longifolia*
Arundo donax	*Brunfelsia pauciflora* 'Floribunda'	*Acalypha wilkesiana*
Baccharis sarothroides	***Buxus microphylla* var. *japonica***	*Acanthus mollis* 'Oak Leaf'
Baccharis viminea	*Camellia japonica*	*Arbutus unedo* 'Compacta'
Buxus microphylla* var. *japonica	*Coccolobis uvifera*	*Berberis* spp.
Callistemon citrinus	*Cotinus coggygria*	*Berberis* × *mentorensis*
Callistemon rigidus	*Dodonaea viscosa*	*Berberis thunbergii*
Carissa macrocarpa	*Dodonaea viscosa* 'Purpurea'	*Buddleia davidii*
Cassia artemisioides	*Dracaena* spp.	*Buxus sempervirens*
Cassia spp.	*Elaeagnus pungens*	*Buxus* spp.
Ceanothus arboreus	*Escallonia rubra*	*Cephalanthus occidentalis*
Cercocarpus betuloides	***Euonymus japonica* 'Grandifolia'**	*Clivia miniata* 'French Hybrid'
Coleonema spp.	*Forsythia* × *intermedia*	*Cornus mas*
Coprosma spp.	*Hakea suaveolens*	*Cotoneaster congestus*
Cortaderia selloana	*Hibiscus rosa-sinensis* 'Ross Estey'	*Cotoneaster horizontalis*
Cotoneaster lacteus	*Ilex glabra*	*Cytisus* × *praecox* 'Moonlight'
Dietes iridiodes	***Ilex vomitoria***	*Elaeagnus umbellata*
Echium fastuosum		*Euonymus alata*
Elaeagnus angustifolia		
Eriogonum arborescens		

Shrubs, cont.		
High	**Moderate**	**Low**
Eriogonum fasciculatum	*Jacquinia armillaris*	*Euonymus japonica*
Eucalyptus lehmannii	*Juniperus chinensis*	'Silver King'
Euonymus japonica	*Juniperus chinensis*	*Fatsia japonica*
Euonymus japonica **'Grandifolia'**	'Armstrong'	*Feijoa sellowiana*
Euphorbia lactea	*Juniperus chinensis*	*Forsythia* × *intermedia*
Fabiana imbricata	'Kazuka'	'Spring Glory'
Fremontia mexicana	*Juniperus chinensis*	*Gardenia augusta*
Furcraea gigantea	'Pfitzerana'	*Hibiscus rosa-sinensis*
Genista spp	*Juniperus communis*	*Hibiscus rosa-sinensis*
Heteromeles arbutifolia	'Hibernica'	'Brillant'
Hibiscus rosa-sinensis	*Juniperus virginiana*	*Ilex* × *altaclarensis* 'Wilsonii'
'President'	*Lantana camara*	***Ilex cornuta***
Ilex cornuta	***Leucophyllum frutescens***	*Ilex cornuta* 'Burfordii'
Ilex vomitoria	*Ligustrum japonicum*	*Ilex cornuta* 'Dazzler'
Isomeris arborea	*Ligustrum lucidum*	*Ilex opaca*
Juniperus californica	*Myrica pensylvanica*	*Juniperus scopulorum* 'Moffetti'
Juniperus chinensis	***Myrtus communis***	*Lantana* 'Confetti'
'Robusta Green'	*Nerium oleander*	***Leptospermum laevigatum***
Juniperus virginiana 'Skyrocket'	'Cherry Ripe'	*Mahonia aquifolium*
Lavatera assurgentiflora	*Ochrosia elliptica*	*Mahonia aquifolium* 'Compacta'
Leptospermum laevigatum	*Olea africana*	*Mahonia pinnata*
Leucophyllum frutescens	*Philodendron selloum*	*Musa acuminata*
Ligustrum spp.	***Phormium tenax***	*Nandina domestica*
Mahonia nevinii	***Pittosporum tobira***	*Philodendron bipinnatifidum*
Myoporum laetum	*Prunus caroliniana*	*Phormium tenax*
Myrtus communis	*Pyracantha cernato-serrata*	'Atropurpureum'
Myrtus communis 'Compacta'	***Pyracantha cernato-serrata***	*Photinia* × *fraseri*
Nerium oleander	**'Graberi'**	*Photinia serrulata*
Olea europaea 'Montra'	***Pyracantha coccinea***	***Pittosporum tobira***
Phormium tenax	*Rhaphiolepis indica*	*Pittosporum tobira* 'Variegata'
Pittosporum crassifolium	'Enchantress'	*Podocarpus macrophyllus* var.
Pittosporum tobira	*Rhaphiolepis umbellata*	*maki*
Plumbago auriculata	***Rhaphiolepsis indica***	*Punica granatum*
Psidium cattleianum	*Shepherdia argentea*	*Rhamnus alaternus*
Punica granatum 'Wonderful'	***Strelitzia reginae***	*Rhododendron* spp.
Pyracantha cernato-serrata **'Graberi'**	***Syzygium paniculatum***	*Rhododendron* spp.
Pyracantha coccinea	*Teucrium chamadrys*	*Ribes* spp.
Rhamnus californica	*Thuja occidentalis*	*Rosa* 'Grenoble'
Rhaphiolepsis indica	*Thuja orientalis*	*Rosa multiflora*
Rhus integrifolia	*Viburnum tinus* 'Robustum'	*Rosa* spp.
Rhus ovata	*Woodwardia* spp.	*Salix purpurea* var. *nana*
Rhus trilobata	*Xylosma congestum*	*Sambucus nigra*
Rosa hybrida	*Zanthoxylum fagara*	*Spiraea vanhouttei*
Rosa rugosa		***Strelitzia reginae***
Simmondsia chinensis		*Viburnum tinus*
Spartium junceum		***Yucca filamentosa***
Syzygium paniculatum		

TABLE 5.8. Salt tolerance of selected landscape plants, cont.

High	Moderate	Low
	Shrubs, cont.	
Tamarix aphylla		
Tamarix gallica		
Tamarix parviflora		
Thuja orientalis 'Aurea'		
Yucca aloifolia		
Yucca filamentosa		
Yucca spp.		
	Trees	
Acacia greggii	*Acacia farnesiana*	*Abies balsamea*
Acacia melanoxylon	***Acacia melanoxylon***	*Acer macrophyllum*
Acer buergerianum	*Acacia* spp.	***Acer pseudoplatanus***
Acer campestre	*Acer platanoides*	*Acer rubrum*
Acer pseudoplatanus	***Araucaria heterophylla***	*Acer saccharinum*
Albizia lophantha	***Arbutus unedo***	*Aesculus* spp.
(syn. *A. distachya*)	***Bauhinia purpurea***	*Albizia julibrissin*
Araucaria heterophylla	***Callistemon viminalis***	*Alnus incana*
Arbutus unedo	*Casuarina*	*Alnus rhombifolia*
Butia capitata	***Ceanothus thyrsiflorus***	*Araucaria araucana*
Callistemon viminalis	*Ceratonia siliqua*	***Bauhinia purpurea***
Caragana arborescens	*Chrysobalanus icaco*	*Betula nigra*
Casuarina cunninghamiana	***Cupressus sempervirens***	*Betula papyrifera*
Casuarina stricta 'Glauca'	*Eucalyptus pulverulenta*	*Betula pendula*
Ceanothus thyrsiflorus	***Eucalyptus sideroxylon***	*Calocedrus decurrens*
Cedrus deodara	*Eucalyptus tereticornis*	*Carpinus betulus*
Cercidium floridum	*Ficus carica*	*Carpinus* spp.
Cercidium spp.	***Fraxinus pennsylvanica***	*Carya illinoensis*
Cercis occidentalis	**var. *lanceolata***	*Carya* spp.
Cinnamomum camphora	*Kopsia arborea*	*Catalpa* spp.
Coccoloba floridana	***Liquidambar styraciflua***	*Cedrus atlantica*
X Cupressocyparis leylandii	*Maclura pomifera*	***Cedrus deodara***
Cupressus sempervirens	***Magnolia grandiflora***	*Celtis australis*
Erythea armata	*Melaleuca quinquenervia*	*Celtis sinensis*
Eucalyptus camaldulensis	*Morus alba* 'Fruitless'	*Cercis* spp.
Eucalyptus citriodora	*Myoporum laetum*	***Cinnamomum***
Eucalyptus cladocalyx	***Olea europaea***	***camphora***
Eucalyptus ficifolia	*Pinus eldarica*	***Citrus* spp.**
Eucalyptus globulus 'Compacta'	*Pinus elliottii*	*Corylus avellana*
Eucalyptus gunnii	*Pinus ponderosa*	*Crataegus × lavallei*
Eucalyptus microtheca	*Pinus sylvestris*	*Crataegus phaenopyrum*
Eucalyptus polyanthemos	*Pinus taeda*	*Cupressus forbesii*
Eucalyptus rudis	*Pinus thunbergiana*	*Cupressus macrocarpa*
Eucalyptus sargentii	***Pistacia chinensis***	*Diospyros virginiana*
Eucalyptus sideroxylon	***Pittosporum***	*Eriobotrya deflexa*
Eucalyptus torquata	***phillyraeoides***	*Eriobotrya japonica*
Euphorbia tirucalis	***Platycladus orientalis***	***Eucalyptus ficifolia***

Trees, cont.

High	Moderate	Low
Ficus microcarpa	*Plumeria rubra*	*Fagus sylvatica*
Fraxinus pennsylvanica* var. *lanceolata	*Populus deltoides*	*Ficus benjamina*
Fraxinus velutina	***Populus fremontii***	*Ficus rubiginosa*
Fraxinus velutina var. *glabra* 'Modesto'	*Prunus* × *blireiana*	*Fraxinus americana*
Gleditsia triacanthos f. *inermis*	*Prunus cerasifera*	*Fraxinus oxycarpa* 'Raywood'
Grevillea banksii	*Punica granatum*	*Ginkgo biloba*
Grevillea 'Canberra'	*Quercus agrifolia*	*Gleditsia triacanthos* 'Moraine'
Grevillea robusta	*Quercus marilandica*	*Juglans nigra*
Hyophorbe spp.	***Quercus palustris***	*Juglans regia*
Jacaranda acutifolia	*Salix alba* 'Vitellina'	***Koelreutaria paniculata***
Koelreutaria paniculata		*Lagerstroemia indica*
Lagunaria patersonii		***Liquidambar styraciflua***
Melaleuca armillaris		*Liriodendron tulipifera*
Melaleuca nesophila		***Magnolia grandiflora***
Metrosideros excelsus		*Magnolia* spp. (deciduous)
Olea europaea		*Malus sylvestris*
Parkinsonia aculeata		*Ostrya virginiana*
Persea americana		***Persea americana***
Pinus brutia		*Picea abies*
Pinus cembroides		*Picea glauca*
Pinus halepensis		*Picea pungens*
Pinus nigra		*Pinus resinosa*
Pinus pinea		*Pinus strobus*
Pinus radiata		***Pistacia chinensis***
Pittosporum phillyraeoides		*Platanus occidentalis*
Platanus × *acerifolia*		***Platycladus orientalis***
Platanus racemosa		***Podocarpus macrophyllus***
Podocarpus macrophyllus		*Populus nigra* 'Italica'
Populus × *canadensis*		*Prunus armeniaca*
Populus alba		*Prunus avium*
Populus fremontii		*Prunus besseyi*
Prunus ilicifolia		*Prunus domestica*
Prunus lyonii		*Prunus dulcis*
Prunus sargentii		*Prunus persica*
Pyrus calleryana		*Prunus tomentosa*
Pyrus kawakamii		*Pseudotsuga menziesii*
Quercus robur		*Pyrus communis*
Quercus rubra		*Quercus bicolor*
Quercus virginiana		***Quercus palustris***
Robinia pseudoacacia		*Sorbus aucuparia*
Salix babylonica		*Tilia* × *euchlora*
Sapium sebiferum		*Tilia americana* 'Redmond'
Schinus molle		*Tilia cordata*
Schinus terebinthifolius		*Tilia* spp.
Sophora japonica		*Tsuga canadensis*
Taxodium distichum		

TABLE 5.8. Salt tolerance of selected landscape plants, cont.

High	Moderate	Low
	Trees, cont.	
Ulmus carpinifolia		*Ulmus alata*
Ulmus glabra		*Ulmus americana*
Ulmus pumila		*Ulmus crassifolia*
Zizyphus jujuba		
	Palms	
Chamaerops humilis	***Arecastrum romanzoffianum***	***Arecastrum romanzoffianum***
Coccothrinax spp.	*Brahea edulis*	*Cordyline australis*
Cocos nucifera	*Coccothrinax argentata*	*Phoenix loureiri*
Cordyline indivisa	*Livistonia chinensis*	***Washingtonia robusta***
Phoenix canariensis	*Sabal palmetto*	
Phoenix dactylifera	*Trachycarpus fortunei*	
Serenoa repens	***Washingtonia filifera***	
Thrinax microcarpa	***Washingtonia robusta***	
Thrinax parviflora	*Zamia* spp.	
Washingtonia filifera		
	Grasses	
Cynodon dactylon	*Bouteloua gracilis*	*Agrostis tenuis*
Paspalum vaginatum	*Dactylis glomerata*	*Eremochloa ophiuroides*
Puccinellia spp.	*Festuca arundinacea*	*Festuca ovina* var. *glauca*
Stenotaphrum secundatum	*Festuca rubra* var. *commutata*	*Festuca rubra*
Uniola paniculata	*Lolium perenne*	*Poa annua*
Zoysia tenuifolia		*Poa pratensis*
		Poa trivialis

Source: After Matheny and Clark 1998.
Note: Boldface type indicates plants listed in more than one category.

TABLE 5.9. Boron tolerance of selected landscape plants (see table 5.7 for common names and references)

High	Moderate	Low
	Groundcovers	
Baccharis pilularis consanguinia	***Rosmarinus officinalis***	***Rosmarinus officinalis***
Ceanothus gloriosus		
Gazania spp.		
Lantana montevidensis		
Rosmarinus officinalis		
	Herbaceous Plants	
Lathyrus odoratus	*Calendula officinalis*	*Delphinium* spp.
Oxalis bowiei	*Callistephus chinensis*	*Helianthus annus*
	Eschscholzia californica	*Lupinus hortwegi*
	Euphorbia pulcherrima 'Barbara Ecke'	*Most ferns*
	Gladiolus spp.	*Pelargonium × hortorum*
	Pelargonium peltatum	*Pelargonium domesticum*
		Viola ordorata
		Viola tricolor
		Zinnia elegans
	Vines	
Bougainvillea spectabilis		
	Shrubs	
Buxus microphylla var. *japonica*	*Abelia × grandiflora*	*Cornus capitata*
Callistemon citrinus	*Gardenia augusta*	*Elaeagnus pungens*
Carissa macrocarpa	***Leucophyllum frutescens***	*Euonymus japonica*
Cassia spp.	***Ligustrum lucidum***	*Euonymus japonica* 'Grandifolia'
Ceanothus arboreus	*Myrtus communis*	*Feijoa sellowiana*
Ceanothus spp.	*Podocarpus macrophyllus* var. *maki*	*Juniperus communis* 'Hibernica'
Ilex cornuta 'Burfordii'	*Rosa chinensis* 'Gloire des Rosomanes'	*Lantana camara*
Cortaderia selloana	*Strelitzia reginae*	***Ligustrum japonicum***
Echium fastuosum	*Syzygium paniculatum*	*Mahonia aquifolium*
Eucalyptus lehmannii	*Teucrium chamadrys*	*Photinia × fraseri*
Heteromeles arbutifolia		*Photinia serrulata*
Hibiscus rosa-sinensis		***Pittosporum tobira***
Juniperus chinensis 'Armstrong'		***Pittosporum undulatum***
Juniperus chinensis 'Kazuka'		*Viburnum tinus*
Juniperus chinensis 'Pfitzerana'		*Xylosma congestum*
Leucophyllum frutescens		
Ligustrum japonicum		
Ligustrum lucidum		
Nerium oleander		
Pittosporum tobira		
Pittosporum undulatum		
Prunus caroliniana		
Psidium cattleianum		
Punica granatum 'Wonderful'		
Rhaphiolepis indica 'Enchantress'		

TABLE 5.9. Boron tolerance of selected landscape plants, cont.

Shrubs, cont.		
High	**Moderate**	**Low**
Tamarix aphylla		
Tamarix parviflora		
Trees		
Acacia spp.	*Acacia melanoxylon*	*Acer saccharinum*
Arbutus unedo	*Carpinus betulus*	*Aesculus carnea*
Callistemon viminalis	*Celtis australis*	*Alnus rhombifolia*
Calodendrum capense	*Ceratonia siliqua*	*Bauhinia purpurea*
Ceanothus thyrsiflorus	*Eriobotrya japonica*	*Betula pendula*
Cedrus atlantica	*Eucalyptus ficifolia*	*Carya illinoensis*
Cedrus deodara	***Eucalyptus polyanthemos***	*Catalpa* spp.
Crataegus phaenopyrum	***Eucalyptus sideroxylon***	*Cercis* spp.
Crinodendron patagua	***Juglans regia***	*Citrus* spp.
Cupressus sempervirens	*Lagerstroemia indica*	*Diosporus kaki*
Eucalyptus camaldulensis	*Liquidambar styraciflua*	*Erythrina caffra*
Eucalyptus citriodora	*Liriodendron tulipifera*	*Ficus carica*
Eucalyptus cladocalyx	*Morus alba*	*Fraxinus velutina* var. *glabra* 'Modesto'
Eucalyptus globulus 'Compacta'	*Olea europaea*	*Gleditsia triacanthos* f. *inermis*
Eucalyptus polyanthemos	*Pinus halepensis*	*Juglans nigra*
Eucalyptus rudis	*Pinus pinea*	***Juglans regia***
Eucalyptus sideroxylon	*Pinus thunbergiana*	*Liquidambar styraciflua* 'Festival'
Fraxinus oxycarpa 'Raywood'	*Platanus × acerifolia*	*Magnolia grandiflora*
Fraxinus velutina	*Platycladus orientalis*	*Magnolia* spp. (deciduous)
***Fraxinus velutina* var. *glabra* 'Modesto'**	*Podocarpus macrophyllus*	*Morus alba* 'Fruitless'
Hymenosporum flavum	*Prunus serrulata*	*Persea americana*
Jacaranda acutifolia	*Quercus ilex*	*Pistacia chinensis*
Lagunaria patersonii	*Sophora japonica*	*Prunus avium*
Laurus nobilis	***Ulmus parvifolia***	*Prunus domestica*
Malus spp.	***Ulmus pumila***	*Prunus persica*
Maytenus boaria	*Zelkova serrata*	*Sequoia sempervirens*
Olea europaea		*Ulmus americana*
Parkinsonia aculeata		***Ulmus parvifolia***
Pinus canariensis		***Ulmus pumila***
Pinus radiata		
Platanus racemosa		
Podocarpus gracillior		
Populus spp.		
Prunus cerasifera 'Atropurpurea'		
Prunus ilicifolia		
Prunus lyonii		
Pyrus kawakamii		
Quercus agrifolia		
Quercus lobata		
Quercus suber		
Salix babylonica		
Salix matsudana 'Tortosa'		

Trees, cont.		
High	**Moderate**	**Low**
Sapium sebiferum		
Schinus molle		
Schinus terebinthifolius		
Vitex luscens		
Palms		
Cordyline indivisa		
Phoenix canariensis		
Phoenix dactylifera		

Source: After Matheny and Clark 1998.
Note: Boldface type indicates plants listed in more than one category.

TABLE 5.10. Methodology and criteria used in evaluating salinity and boron tolerance in selected references cited in table 5.7

	Relative Tolerance			
Reference	**Low**	**Moderate**	**High**	**Methodology/criteria**
Salinity of irrigation water: EC_w				
Bernstein 1958	0.75–1.5	1.5–3.0	>3.0	Plots salinized with NaCl and $CaCl_2$ to provide 0.7, 4.4, and 7.8 mmhos/cm in irrigation water. Tolerance evaluated by growth reduction and appearance.
Cal Poly n.d.	—	—	1.0–2.5	Observations at landscapes in San Bernadino County, CA. Plants irrigated with water with conductivity of 1.0–2.5 mmhos/cm.
Farnham, Ayers, and Hasek 1985	0.75–1.5	1.5–3.0	>3.0	Results of work done by Branson and Davis 1965, University of California Cooperative Extension.
Skimina 1980	180–300 mhos × 10^{-5}	400–800	1,000–1,200	Source of salts: 50% fertilizer and 50% NaCl. Plants grown in containers. Soil salinity usually 2–3 times higher than irrigation water. Classification based on not not more than 50% relative growth reduction, no leaf burn, negligible mortality.
Van Arsdel 1980	—	—	—	Ratings based on personal experience with waters containing 48–140 ppm Cl and 188–445 ppm Na.
Wu et al. 1999	—	—	—	Tested potable; low salt (500mg/l NaCl) and high salt (1,500 mg/l NaCl) on plants in loam field soil and containers under sprinkler and drip irrigation.
Salinity of soil: EC_e				
Bernstein 1958	2.0–3.0 mmhos/cm	4.0–6.0	6.0–12.0	No methods stated.
Dirr 1978	—	—	—	Container plants irrigated with 0.25 N NaCl daily. Plants ranked on an appearance index.
Francois 1980	2.0–4.0 dS/m	6.0–8.0	8.0–10.0	Methods not given. Plants with EC_e 6.0–8.0 rated as having good tolerance.
Francois 1982	<4.0	4.0–6.0	7.0–9.0	Maximum root zone salinity without foliar injury. Methods similar to Francois and Clark 1978.

TABLE 5.10. Methodology and criteria used in evaluating salinity and boron tolerance in selected references cited in table 5.7, cont.

Reference	Relative Tolerance: Low	Moderate	High	Methodology/criteria
Francois and Clark 1978	<3.0	3.0–6.0	6.0–9.0	Rating criteria: <50% growth reduction; no leaf injury; plants aesthetically appealing. Experiment run for 3 years in sandy loam soil. Plants irrigated with NaCl and $CaCl_2$ added to yield EC_w of 0.7 (control), 4.4, and 7.8 mmhos/cm. Average EC_e was 1.0, 4.3, and 7.0 mmhos/cm. Soil salinity was uniform with depth throughout the root zone during summer.
Glattstein 1989	—	—	—	No methods stated; authors assumed plants included in list were in "moderate" category.
Harivandi 1988	<4.0 dS/m	4.0–8.0	8.0–16.0	—
Morris and Devitt 1990	—	—	—	Plants grown for 8 years in silty clay loam soil with saline groundwater. EC_w 26.0–40.0 at 3-foot depth. EC_e in root zone 8.0–13.0 mmhos/cm. Authors ranked trees by appearance.
		Boron: ECw		
Eaton 1944	<1.0	5	10.0–25.0	Plants grown from seed, outdoors, in large sand culture. Irrigated with 0.03, 1, 5, 10, 15 and 25 ppm boron.
Farnham, Ayers, and Hasek 1985	0.5–1.0 mg/l	1.0–2.0	2.0–10.0	Adapted from Eaton 1935.
Francois and Clark 1978	0.5	2.5	7.5	Plants grown outdoors in sand culture. Classification based on growth reduction and overall plant appearance.
Questa 1987	—	—	—	Inventoried plants growing in Concord, CA, parks that were irrigated with high-boron water. Species were evaluated for injury and ranked according to sensitivity, based on boron concentration in soil and water, and severity of toxicity symptoms.
San Diego 1963	—	0.75–3.0	—	Observations at landscapes in San Bernadino County, CA, irrigated with boron water; ratings based on growth and leaf injury.

Source: After Matheny and Clark 1998.

Soil salinity affects irrigation management. As the soil dries, the concentration of salts is increased, which increases the potential for toxicity. Keeping soil moist reduces the potential for toxicity. In addition, osmotic tension decreases the availability of water to plants in saline soils. It may be necessary to increase irrigation frequency and/or duration when irrigating with saline water or managing a saline soil.

Care should be taken to avoid heavy applications of fertilizer, as they contribute to soil salinity. Where salinity is of concern, select high-analysis formulations with low salt hazard. Animal manures, mushroom compost, and sewage sludge should be tested for salts before applying to landscapes. If misapplications of fertilizer were made or saline soil amendments applied, leach with good-quality water to move the salts below the root zone.

Salt deposits on foliage can be washed off with good-quality water. When irrigating with saline water, avoid application by sprinklers.

Treating sodic soils requires providing a soluble source of calcium. Gypsum ($CaSO_4 \cdot 2H_2O$) is the material commonly used. The calcium in gypsum displaces sodium in the soil, freeing the sodium to be

Figure 5.56. The lower foliage of this coast live oak *(Quercus agrifolia)* suffered salt damage when it was wetted with saline water through the irrigation system.

leached below the root zone of the plant. Excessive use of gypsum can cause problems, however, so test the soil first to make sure that sodium is excessive. A soil-testing laboratory can determine how much gypsum is required to reclaim sodic soil. If sodic soils are also saline, gypsum should be incorporated before leaching treatments are applied. There is no need to apply gypsum to soils that are simply alkaline (high pH) or calcareous. If the soil is calcareous (containing $CaCO_3$), sulfur can be applied to release the calcium for displacement of sodium. The reaction may take several months to several years (Cardon and Mortvedt 1999). See "Problems Related to pH," below, for more information. A summary of salt-related problems is provided in table 5.11.

TABLE 5.11. Summary of salt-related problems

Problem or condition	Symptoms	Diagnosis	Occurrence and aggravating factors	Look-alike disorders	Treatment
			Soil		
saline	Stunted growth; chlorosis; leaf tip and marginal burn; defoliation; death.	Test soil for EC_e; saline soils have EC_e greater than 4 dS/m. If irrigation water could be the source of salts, test water for total dissolved solids (TDS) and EC_w.	Species sensitivity to salts; low soil moisture; high water table; poor drainage; irrigation with saline water; application of deicing salts; heavy application of fertilizer or saline soil amendment.	Mineral deficiency; drought; herbicide toxicity; wind burn; acute air pollution; high light exposure.	Correct drainage problems; leach with good-quality water; select tolerant plants.
sodic	Stunted growth; cholorsis; necrosis; death. May be white or black crust on soil surface. Water may pond on soil surface.	Test soil for sodium adsorption ratio (SAR) or exchangeable sodium percentage (ESP). Sodic soils have SAR >6 and ESP >10.	Species sensitivity to sodium; high water table; poor drainage; irrigating with water high in sodium; using softened water; application of NaCl as deicing salt.	Mineral deficiency; drought; herbicide toxicity; wind burn; acute air pollution; high light exposure.	Incorporate gypsum, or sulfur in calcerous soils; leach with good-quality water.
saline-sodic	Stunted growth; chlorosis; necrosis; death. May be white or black crust on soil surface. Water may pond on soil surface.	Test soil for EC_e and sodium adsorption ratio (SAR). Saline-sodic soils have EC_e >4.0 dS/m and SAR >6.	Species sensitivity to sodium and salt; high water table; poor drainage; irrigating with water high in sodium and salt; using softened water; application of NaCl as deicing salt; low soil moisture.	Mineral deficiency; drought; herbicide toxicity; wind burn; acute air pollution; high light exposure.	Incorporate gypsum, or sulfur in calcerous soils; leach with good-quality water.
chloride	Stunted growth; necrosis of leaf tips or margins; bronzing; premature yellowing and abscission of leaves; chlorosis.	Test soil and tissue for Cl.	Species sensitivity to salts; low soil moisture; high water table; poor drainage; irrigation with high-Cl water; application of deicing salts; heavy application of chloride-containing fertilizer; close proximity to swimming pool.	Mineral deficiency; drought; herbicide toxicity; wind burn; acute air pollution; high light exposure.	Correct drainage problems; leach with good-quality water; select tolerant plants.

TABLE 5.11. Summary of salt-related problems, cont.

Problem or condition	Symptoms	Diagnosis	Occurrence and aggravating factors	Look-alike disorders	Treatment
			Soil, cont.		
boron	Yellowing of leaf tip, followed by progressive chlorosis and necrosis of margins and between veins; necrosis is black and may appear as small spots near leaf margin.	Test soil or leaves for boron.	Species sensitivity to boron; irrigation with high-boron water; application of certain sewage effluent wastes; application of borate-containing herbicides.	Mineral deficiency; drought; herbicide toxicity; wind burn; acute air pollution; high light exposure.	Correct drainage problems; leach with good-quality water; select tolerant plants.
sodium	Mottled and interveinal chlorosis progressing to necrotic leaf tips, margins, and between veins.	Test soil and/or leaves for sodium.	Species sensitivity to sodium; irrigation with chemically softened water or other water high in sodium; application. of NaCl as deicing salt	Mineral deficiency; drought; herbicide toxicity; wind burn; acute air pollution; high light exposure.	Incorporate gypsum, or sulfur in calcerous soils; leach with good-quality water.
ammonium	Reduced growth; chlorosis; small necrotic spots on leaves.	Test soil for ammonium.	Species sensitivity to ammonium; heavy application of ammonium fertilizer; incorporation of soil amendment high in ammonium.	Mineral deficiency; herbicide toxicity; high light exposure.	Leach with good-quality water.
			Leaves		
deicing salts	Damage occurs on the side of the plant facing the road, and to the splash height. In conifers, needles turn brown from tips downward. In broadleaves and confiers, bud, twig, branch, and whole plant death may occur.	Test foliage for salts.	Species sensitivity; length of exposure and concentration of salts in spray.	Herbicide toxicity.	Wash off foliage. Use less-toxic deicing salt.
sprinkler irrigation	Leaf necrosis; damage occurs on foliage wetted by sprinkler.	Test foliage and water for chloride.	Irrigation water with >100 mg/l Cl; species sensitivity.	Drought; herbicide toxicity; wind burn; acute air pollution.	Wash off foliage with good-quality water.
ocean spray	Foliage necrotic on windward side of plant.	Test foliage for chloride.	Exposure of salt-sensitive species to wind-driven spray.	Mineral deficiency; drought; herbicide toxicity; wind burn; acute air pollution.	Wash off foliage with good-quality water.

Problems Related to pH

Soil reaction, expressed as pH, refers to the acidity or alkalinity of a soil, that is, the relative concentrations of hydrogen (acid) and hydroxide (alkaline) ions. A logarithmic-based scale from 0 to 14 is used to express pH. Equal concentrations of hydrogen and hydroxide ions produce a neutral reaction, a pH of 7.0. As a soil becomes more acidic, its pH decreases; as it becomes more alkaline, its pH increases. Soil pH

- affects the availability of plant nutrients in the soil
- influences the solubility of certain elements in the soil (e.g., aluminum, manganese) that may become toxic
- affects the population and activity of soil microorganisms
- has a direct effect on the root cell function, which can influence water and nutrient uptake

Under most conditions, soil pH has a primarily nutritional effect on plants. An optimal pH for the availability of nutrients essential for plant growth, without becoming toxic, is from 5.5 to 7.0. At pH levels above or below this narrow range, the availability of one or more nutrient elements can be significantly reduced or increased. In strongly acid soils (below pH 5.5), the availability of calcium, magnesium, phosphorus, nitrogen, sulfur, molybdenum, and boron is reduced. At the same time, the availability of aluminum, iron, manganese, zinc, and copper is increased and could be toxic to plants and microorganisms. Except for molybdenum and chloride, micronutrients become less available as soil alkalinity increases. In strongly alkaline soils (above pH 7.5), iron, manganese, zinc, and copper become unavailable for plant use. For more information on mineral deficiencies, see "Nutrient Deficiencies," p. 71.

Microbial activity is also affected by pH. While fungi function over a wide pH range, bacteria and actinomycetes are favored by a slightly acid to alkaline pH. Nitrifying bacteria are inhibited when the pH is less than 5.5, and they are absent in very acid soils (Craul 1992).

Interactions with Aeration

The interaction between soil pH and aeration may affect micronutrient availability. Iron, manganese, and copper are generally more available in poorly drained or flooded acidic soils and may reach toxic levels (Brady and Weil 1996).

Calcareous Soils

Calcareous, or high-lime, soils contain calcium carbonate ($CaCO_3$). Calcareous soils are typically light in color, silty in texture, and often poorly drained. The pH is alkaline, usually ranging from 7.5 to 8.5.

Symptoms

Because soil pH affects the availability of nutrient elements, primary symptoms are seen as nutrient excesses or deficiencies in sensitive plants. Common symptoms are described in table 5.12.

Acid pH

Three micronutrients may become toxic at low pH: aluminum, manganese, and copper. Aluminum toxicity that occurs at pH less than 4.5 cannot be diagnosed from visual symptoms or the aluminum content of foliage (Chapman 1965). Roots typically are discolored, short, and stubby, but this may not be distinguishable on plants grown in some soils. Growth is reduced, but this in itself is not sufficient for diagnosis. Stunted roots have difficulty absorbing immobile nutrients such as phosphorus. Because of poor root growth and the reactions between aluminum and phosphorus, aluminum toxicity in plants resembles phosphorus deficiency (Singer and Munns 1987). Figure 5.57 illustrates the differences in plant growth in very acidic and slightly acidic soil.

Manganese toxicity produces varied patterns of leaf distortion, yellowing, and necrosis, depending on the plant species. In citrus, leaves develop marginal yellowing and tiny necrotic spots (Labanauskas 1966).

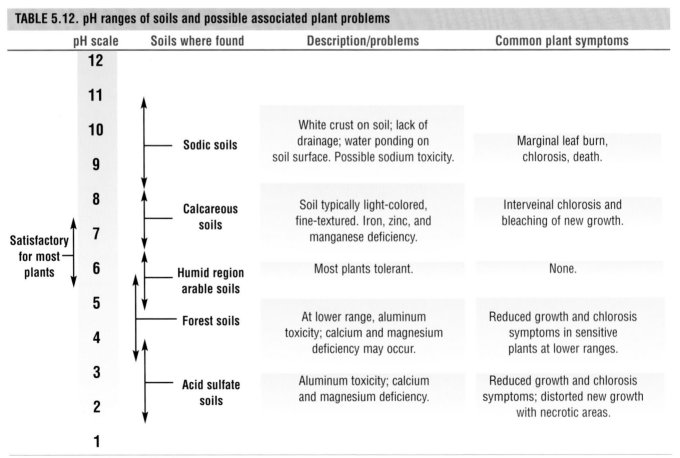

Figure 5.57. The pH on this cut slope ranged from 3.2, where few plants grew (left side) to 6.2, where growth was normal (right side).

Copper toxicity is expressed first as reduced growth. It may cause iron chlorosis symptoms by depressing the iron concentration in leaves. Like aluminum toxicity, it is associated with stunting, reduced branching, and thickening and discoloration of roots in many plants (Reuther and Labanauskas 1966).

Alkaline pH

The primary symptoms associated with alkaline pH are deficiencies in iron, zinc, and manganese. These deficiencies are expressed as chlorosis of new growth (fig. 5.58). The patterns of chlorosis vary slightly. In broadleaf shrubs and trees, iron deficiency causes interveinal chlorosis with narrow bands of green along the veins. With manganese deficiency, the bands of green are typically wider. Zinc deficiency causes a more mottled chlorosis pattern and abnormally small leaves and short internodes. It is often difficult to distinguish among these

TABLE 5.12. pH ranges of soils and possible associated plant problems

pH scale	Soils where found	Description/problems	Common plant symptoms
12–9	Sodic soils	White crust on soil; lack of drainage; water ponding on soil surface. Possible sodium toxicity.	Marginal leaf burn, chlorosis, death.
8–7	Calcareous soils	Soil typically light-colored, fine-textured. Iron, zinc, and manganese deficiency.	Interveinal chlorosis and bleaching of new growth.
6–5	Humid region arable soils	Most plants tolerant.	None.
5–4	Forest soils	At lower range, aluminum toxicity; calcium and magnesium deficiency may occur.	Reduced growth and chlorosis symptoms in sensitive plants at lower ranges.
3–2	Acid sulfate soils	Aluminum toxicity; calcium and magnesium deficiency.	Reduced growth and chlorosis symptoms; distorted new growth with necrotic areas.

Satisfactory for most plants: 7–6

Source: Adapted from Brady and Weil 1996.

Figure 5.58. A common symptom of plants growing in alkaline soil is interveinal chlorosis on young foliage. This is due to unavailability of iron, zinc, or manganese at alkaline pH.

elemental deficiencies because their symptoms are so similar. To further complicate diagnosis, more than one element may be deficient; a range in symptoms may result. Diagnosis is best made with tissue analysis (see "Nutrient Deficiencies," p. 71).

Sensitive plants growing in calcareous soil are often deficient in iron, zinc, or manganese because the ions are oxidized and unavailable for plants (fig. 5.59). New growth may be completely chlorotic, with necrotic areas.

Occurrence

Soil pH is determined by the parent material from which the soil is derived, amount of rainfall, vegetation, and soil drainage. Acidic soils tend to develop in high-rainfall and high-humidity areas, and also under coniferous forests. Alkaline soils tend to develop in low rainfall and poorly drained areas, and also under grasses.

Most soils in California are slightly acidic to moderately alkaline. Calcareous conditions are common in flood plains and valleys, and low pH soils are occasionally encountered (table 5.13). Acid subsoils may be exposed when soil cuts are made during

Figure 5.59. Common symptoms of sensitive plants growing in calcareous soil includes poor growth (A), interveinal chlorosis (B) on mock strawberry *(Duchesnea indica)*, and bleaching (C) on photinia *(Photinia fraseri)*.

site grading. Landscape soils are often highly manipulated and may differ from the native soil.

Look-Alike Disorders

Calcium deficiency, copper toxicity, and herbicide toxicity (e.g., caused by Surflan) may cause stunted root symptoms similar to low pH. Plants may appear deficient in phosphorus due to the inability of stunted roots to absorb phosphorus. Symptoms of iron, zinc, and manganese deficiencies can be caused by a number of factors (see "Nutrient Deficiencies," p. 71).

Diagnosis

Problems caused by soil pH can be diagnosed by collecting a soil sample from the root area and testing it for pH. Tests should include the related factors of calcium carbonate concentration (percent lime), as its presence affects possible treatments and sensitive plant species. Whether lime is present is more important to diagnosing problems than the quantity of lime in a soil. The amount of calcium carbonate indicates the buffering capacity of the soil against acidification should such treatment be considered (see "Remedies," p. 121). Tissue analyses can help diagnose suspected mineral toxicities that may occur at low pH (table 5.14).

Figure 5.60. Plants vary in their tolerance to unusually low or high pH. This soil has a pH of 4.5. Gazania (*Gazania* sp.) on the left appears normal, while California poppy (*Eschscholzia californica*) on the right is chlorotic and stunted.

Sensitive and Tolerant Species

Plants vary considerably in their tolerance to acid and alkaline conditions, although most plants grow satisfactorily in a pH range of 5.5 to 7.0 (table 5.15; fig. 5.60). In general, species are tolerant of the soils in which they have evolved; pH requirements can be estimated by evaluating the soils

TABLE 5.13. Occurrence of acidic, alkaline, and calcareous soils in California

Soil condition	Land type	Location
calcareous	alluvial fan and flood plain soils of desert region	Imperial Valley Palo Verde Valley parts of Mojave Desert southwest portion of San Joaquin Valley
saline and sodic	imperfectly drained basin soils	parts of San Joaquin Valey Surprise Valley (Modoc Co.) Honey Lake Valley (Lassen Co.) playas of the Mojave Desert region
slightly to moderately acidic	terrace land valley basin organic soils mountains	coast from Del Norte Co. to San Luis Obispo Co. Sacramento–San Joaquin Delta Sierra Nevada mountain range and other high-elevation mountains of Northern and Southern California
strongly acidic	coastal range	subsoils exposed in soil cuts in Southern California (Capistrano formation)

TABLE 5.14. Evaluating toxicity from soil and tissue analyses

Element	pH	Soil analysis: Not harmful	Soil analysis: Probably toxic	Soil analysis: High	Tissue analysis: Normal	Tissue analysis: Toxic
aluminum	<4.5	<0.5 ppm	0.5–1.0 ppm	>1.0 ppm	—	—
copper	<5.0	—	—	>150 ppm	—	>20 ppm
manganese	<5.5	—	—	—	—	>1,000 ppm

Source: Chapman 1965.

from the region in which the plants are native. Most pines are generally tolerant to highly acid soils. Spruce grow best on less acid soils. Hardwoods, on the other hand, generally tolerate alkaline soils.

Acid-loving plants, most of which are native to acidic soils, have difficulty absorbing iron and require a high degree of soluble iron in the soil. They perform best in acidic soils with pH of 5; at higher pH they tend to show iron, zinc, or manganese deficiency symptoms (fig. 5.61). Plants that grow poorly in very acid soils (pH less than 5) are usually affected by aluminum toxicity.

Soils that are calcareous may limit growth and cause iron deficiency symptoms in species that are normally tolerant of low to moderate alkalinity.

Remedies

Although rarely a problem in California, acid soils can be limed to increase the pH. The amount of lime required depends on the pH, the soil texture, and the cation exchange capacity of the soil.

Soil pH is usually reduced by incorporating acidifying materials such as sulfur, or gypsum, which contains sulfur. When sulfur is added to the soil, bacteria convert it to sulfuric acid. Warm temperatures, moist soil, and oxygen are required for bacterial activity. The reaction takes place slowly over 6 to 8 weeks under optimal conditions.

Calcium carbonate in the soil acts as a buffer against lowering the pH. If sulfur is applied to calcareous soils, the calcium carbonate reacts with sulfuric acid to produce water, carbon dioxide, and calcium sulfate. No increase in hydrogen ion concentration occurs, so there is no change in soil pH. The pH of calcareous soils and soils irrigated with water containing calcium carbonate cannot reasonably be lowered. There is no remedy for high-pH calcareous soils; tolerant species should be planted in them. Symptoms of micronutrient deficiency can be treated with foliar applications of appropriate elements (see "Nutritient Deficiencies," p. 71). Soil pH problems are summarized in table 5.16.

Figure 5.61. Pin oak *(Quercus palustris)* requires acidic soils. When grown in alkaline soil (pH 7.8 in this photo), foliage shows symptoms of reduced growth, chlorosis, and necrosis (normal on left).

TABLE 5.15. pH tolerance of selected landscape plants

Scientific name	Common name	pH tolerance* Acid	Alkaline	Alkali	Source†
		Trees			
Acacia longifolia	Sydney golden wattle			x	Perry 1989
Acacia melanoxylon	blackwood acacia		x	x	Hoyt 1978; Perry 1989; Svihra and Coate 1996
Acer campestre	hedge maple		x		Kuhns and Rupp 2000
Acer negundo	box elder		x		Kuhns and Rupp 2000
Acer nigrum	black maple		x		Kuhns and Rupp 2000
Acer platanoides	Norway maple		x		Kuhns and Rupp 2000
Acer pseudoplatanus	sycamore maple		x		Kuhns and Rupp 2000
Acer rubrum	red maple	x			Wysong et al. 2000
Acer saccharinum	silver maple	x			Wysong et al. 2000
Acer saccharum	sugar maple	x	x		Kuhns and Rupp 2000; Wysong et al. 2000
Acer saccharum grandidentatum	big-tooth maple		x		Kuhns and Rupp 2000
Agonis flexuosa	peppermint tree		x		Hoyt 1978; Perry 1989; EBMUD 1990
Ailanthus altissima	tree-of-heaven		x	x	Hoyt 1978; Kuhns and Rupp 2000
Albizzia distachya	plume albizia			x	Hoyt 1978
Annona cherimola	cherimoya		x		Hoyt 1978
Brachychiton populneus	bottle tree			x	Hoyt 1978; Perry 1989
Calocedrus decurrens	incense cedar		x		Kuhns and Rupp 2000
Casuarina cunninghamiana	river she-oak		x		Svihra and Coate 1996
Casuarina stricta	coast beefwood		x		Svihra and Coate 1996
Catalpa bignonioides	common catalpa	x	x		Kuhns and Rupp 2000; Wysong et al. 2000
Catalpa speciosa	western catalpa	x	x		Kuhns and Rupp 2000; Wysong et al. 2000
Cedrus atlantica	atlas cedar		x		EBMUD 1990
Cedrus deodara	deodar cedar		x		Svihra and Coate 1996
Celtis australis	European hackberry			x	Hoyt 1978; EBMUD 1990
Celtis occidentalis	common hackberry		x		Kuhns and Rupp 2000
Cercidium floridum	blue palo verde		x		Perry 1989
Cercis canadensis	eastern redbud		x		Kuhns and Rupp 2000
Cercis occidentalis	western redbud		x		Kuhns and Rupp 2000
Cercocarpus ledifiorius	curl-leaf mountain mahogany		x		Kuhns and Rupp 2000
Chilopsis linearis	desert willow		x	x	Hoyt 1978; Kuhns and Rupp 2000
Cinnamomum camphora	camphor tree			x	Hoyt 1978
Cladrastis lutea	yellow wood		x		Kuhns and Rupp 2000
Cornus florida	flowering dogwood	x			Wysong et al. 2000
Corylus colurna	Turkish hazel		x		Kuhns and Rupp 2000
Corylus cornuta	western hazelnut		x		Kuhns and Rupp 2000
Cotinus obovatus	American smoketree		x		Kuhns and Rupp 2000
Crataegus × lavallei	Carriere hawthorn		x		Kuhns and Rupp 2000
Crataegus crus-galli	cockspur thorn		x		Kuhns and Rupp 2000
Crataegus phaenopyrum	Washington thorn		x		Kuhns and Rupp 2000
Crataegus viridis	green hawthorn		x		Kuhns and Rupp 2000
Cupressus arizonica	Arizona cypress		x		EBMUD 1990; Svihra and Coate 1996; Kuhns and Rupp 2000
Cupressus sempervirens	Italian cypress		x		Kuhns and Rupp 2000

Scientific name	Common name	pH tolerance* Acid	Alkaline	Alkali	Source†
		Trees, cont.			
Cupressus spp.	cypress		x		Perry 1989
Diospyros kaki	Japanese persimmon		x		Hoyt 1978
Embothrium coccineum	Chilean fire bush	x			Hoyt 1978
Eriobotrya japonica	Japanese loquat		x		EBMUD 1990; Kuhns and Rupp 2000
Eriobotrya spp.	loquat		x		Hoyt 1978; Perry 1989
Eucalyptus camadulensis	red gum		x		EBMUD 1990; Perry 1989; Svihra and Coate 1996
Eucalyptus cladocalyx	sugar gum		x		Perry 1989
Eucalyptus globulus	Tasmanian blue gum		x		Perry 1989
Eucalyptus leucoxylon rosea	pink flowered white ironbark	x			EBMUD 1990; Perry 1989
Eucalyptus microtheca	coolibah		x		EBMUD 1990
Eucalyptus polyanthemos	silver dollar gum		x		Perry 1989
Eucalyptus rudis	flooded gum		x		Perry 1989
Eucalytpus saligna	Sydney blue gum		x		Svihra and Coate 1996
Ficus macrophylla	Moreton Bay fig			x	Hoyt 1978
Fraxinus americana	white ash		x		Kuhns and Rupp 2000
Fraxinus excelsior	European ash		x		Kuhns and Rupp 2000
Fraxinus oxycarpa 'Raywood'	Raywood ash		x		Svihra and Coate 1996
Fraxinus pennsylvanica	green ash		x		Kuhns and Rupp 2000
Fraxinus uhdei	evergreen ash		x		EBMUD 1990
Fraxinus velutina	Arizona ash			x	Hoyt 1978
Ginkgo biloba	maidenhair tree	x	x		Kuhns and Rupp 2000; Wysong et al. 2000
Gleditsia triacanthos	honey locust		x		Kuhns and Rupp 2000
Grevillea robusta	silk oak			x	Hoyt 1978
Gymnocladus dioicus	Kentucky coffee tree		x		Kuhns and Rupp 2000
Juglans cinerea	butternut		x		Kuhns and Rupp 2000
Juglans nigra	black walnut		x		Kuhns and Rupp 2000
Juglans regia	English walnut		x		Kuhns and Rupp 2000
Juglans spp.	walnut		x		Hoyt 1978
Juniperus spp.	juniper		x		Hoyt 1978
Juniperus virginiana	eastern red cedar		x		Kuhns and Rupp 2000
Koelreuteria paniculata	goldenrain tree		x	x	Hoyt 1978; EBMUD 1990; Kuhns and Rupp 2000
Laburnum × *watereri*	goldenchain tree		x		Kuhns and Rupp 2000
Lagerstroemia indica	crape myrtle		x		Kuhns and Rupp 2000
Lagunaria patersonii	primrose tree			x	Hoyt 1978
Laurus 'Saratoga'	Saratoga bay		x		EBMUD 1990
Laurus nobilis	Grecian laurel		x		Perry 1989
Liquidambar spp.	sweetgum	x			Hoyt 1978
Liriodendron tulipifera	tulip tree	x			Wysong et al. 2000
Litchi chinensis	litchi nut	x			Hoyt 1978
Lophostemon confertus	Brisbane box		x		Svihra and Coate 1996
Maclura pomifera	Osage orange		x	x	Hoyt 1978; Kuhns and Rupp 2000
Magnolia grandiflora	southern magnolia	x			Hoyt 1978
Malus spp.	crabapple		x		Kuhns and Rupp 2000
Melaleuca quinquenervia	cajeput tree		x	x	EBMUD 1990; Svihra and Coate 1996; Hoyt 1978; Perry 1989

TABLE 5.15. pH tolerance of selected landscape plants, cont.

Scientific name	Common name	pH tolerance* Acid	Alkaline	Alkali	Source†
		Trees, cont.			
Melaleuca styphelioides	black tea tree		x	x	Hoyt 1978; Perry 1989; Svihra and Coate 1996
Melia azedarach	chinaberry		x	x	Hoyt 1978; Kuhns and Rupp 2000; Svihra and Coate 1996
Morus alba	white mulberry		x		Kuhns and Rupp 2000
Morus rubra	red mulberry			x	Hoyt 1978; Kuhns and Rupp 2000
Olea europaea	European olive		x		Hoyt 1978; Perry 1989; Svihra and Coate 1996
Parkinsonia aculeata	Mexican palo verde		x	x	Hoyt 1978; Perry 1989
Persea thunbergii	avocado		x		Hoyt 1978
Phoenix spp.	date palm			x	Hoyt 1978
Pinus edulis	pinon		x		Kuhns and Rupp 2000
Pinus eldarica	Afghan pine		x		EBMUD 1990
Pinus halepensis	Aleppo pine		x	x	Hoyt 1978; Perry 1989; Kuhns and Rupp 2000
Pinus monophylla	singleleaf piñon pine		x		Kuhns and Rupp 2000
Pinus mugo	mugho pine		x		Kuhns and Rupp 2000
Pinus nigra	Austrian black pine		x		Kuhns and Rupp 2000
Pinus pinea	Italian stone pine		x		Svihra and Coate 1996
Pinus sabiniana	foothill pine		x		Svihra and Coate 1996
Pinus sylvestris	Scotch pine		x		EBMUD 1990
Pinus thunbergiana	Japanese black pine		x		Svihra and Coate 1996
Pistacia chinensis	Chinese pistache		x	x	Hoyt 1978; Kuhns and Rupp 2000
Pittosporum phillyraeoides	willow pittosporum			x	Perry 1989
Pittosporum spp.	pittosporum		x		Hoyt 1978
Platanus occidentalis	American sycamore		x		Kuhns and Rupp 2000
Platanus racemosa	California sycamore			x	Hoyt 1978
Podocarpus macrophyllus	yew pine		x		Svihra and Coate 1996
Populus × canadensis	Carolina poplar		x		Kuhns and Rupp 2000
Populus alba	white poplar		x		Kuhns and Rupp 2000
Populus fremonti	western cottonwood		x	x	Hoyt 1978; Kuhns and Rupp 2000; Svihra and Coate 1996
Populus nigra 'Italica'	Lombardy poplar		x		Kuhns and Rupp 2000
Populus trichocarpa	black cottonwood		x		Kuhns and Rupp 2000
Prosopis glandulosa	honey mesquite		x		Kuhns and Rupp 2000
Pseudotsuga menziesii	Douglas fir	x	x		Kuhns and Rupp 2000; Wysong 2000
Pyrus calleryana cvs.	callery pear		x		EBMUD 1990; Kuhns and Rupp 2000; Svihra and Coate 1996
Pyrus communis	pear		x		Kuhns and Rupp 2000
Quercus agrifolia	coast live oak			x	Hoyt 1978
Quercus alba	white oak	x			Wysong et al. 2000
Quercus cerris	turkey oak		x		Kuhns and Rupp 2000
Quercus chrysolepis	canyon live oak	x			Hoyt 1978
Quercus coccinea	scarlet oak	x			Wysong et al. 2000
Quercus gambelii	Gambell oak		x		Kuhns and Rupp 2000
Quercus ilex	holly oak		x		Hoyt 1978; EBMUD 1990; Svihra and Coate 1996

Scientific name	Common name	pH tolerance* Acid	Alkaline	Alkali	Source†
		Trees, cont.			
Quercus lobata	valley oak			x	Hoyt 1978
Quercus macrocarpa	bur oak	x	x		Kuhns and Rupp 2000; Wysong et al. 2000
Quercus muehlenbergii	chinquapin oak		x		Kuhns and Rupp 2000
Quercus palustris	pin oak	x			Wysong et al. 2000
Quercus robur	English oak		x		Kuhns and Rupp 2000
Quercus rubra	northern red oak	x			Wysong et al. 2000
Quercus shumardii	Shumard red oak		x		Kuhns and Rupp 2000
Quercus suber	cork oak			x	Hoyt 1978
Quercus virginiana	southern live oak	x			Hoyt 1978
Rhus lancea	African sumac		x		Svihra and Coate 1996
Robinia × *ambigua*	Idaho locust		x		Kuhns and Rupp 2000
Robinia pseudoacacia	black locust			x	Hoyt 1978; Kuhns and Rupp 2000
Salix babylonica	weeping willow		x		Kuhns and Rupp 2000
Salix matsudana	Hankow willow		x		Kuhns and Rupp 2000
Salix nigra	black willow		x		Kuhns and Rupp 2000
Schinus molle	California pepper tree		x	x	Hoyt 1978; EBMUD 1990; Svihra and Coate 1996
Schinus spp.	peppertree		x		Hoyt 1978; Perry 1989
Sequoia sempervirens	coast redwood	x			Hoyt 1978
Sequoiadendron giganteum	giant sequoia	x			Hoyt 1978
Sophora japonica	Japanese pagoda tree		x		Kuhns and Rupp 2000
Stenocarpus sinuatus	firewheel tree	x			Hoyt 1978
Syringa reticulata	Japanese tree lilac		x		Kuhns and Rupp 2000
Tamarix aphylla	athel tree			x	Hoyt 1978
Taxodium distichum	bald cypress	x			Hoyt 1978
Taxodium mucronatum	Montezuma cypress		x		Hoyt 1978
Thuja occidentalis	American arborvitae		x		Kuhns and Rupp 2000
Thuja orientalis	oriental arborvitae		x		Kuhns and Rupp 2000
Thuja plicata	western red cedar		x		Kuhns and Rupp 2000
Tilia × *euchlora*	Crimean linden		x		Kuhns and Rupp 2000
Tilia americana	American linden		x		Kuhns and Rupp 2000
Tilia cordata	little-leaf linden		x		Kuhns and Rupp 2000
Tilia tomentosa	silver linden		x		Kuhns and Rupp 2000
Tipuana tipu	tipu tree		x		Hoyt 1978
Tristania conferta (see Lophostemon)					
Ulmus americana	American elm		x		Kuhns and Rupp 2000
Ulmus parvifolia	Chinese evergreen elm		x		Kuhns and Rupp 2000
Ulmus pumila	Siberian elm			x	Hoyt 1978; Kuhns and Rupp 2000
Washingtonia filifera	California fan palm			x	Hoyt 1978
Yucca brevifolia	Joshua tree		x	x	Hoyt 1978; Kuhns and Rupp 2000
Ziziphus jujuba	jujube		x	x	Hoyt 1978; EBMUD 1990; Kuhns and Rupp 2000
		Shrubs			
Abelia floribunda	Mexican abelia	x			Hoyt 1978
Abelia grandiflora	glossy abelia		x		Svihra and Coate 1996
Acacia armata	kangaroo thorn			x	Hoyt 1978

TABLE 5.15. pH tolerance of selected landscape plants, cont.

Scientific name	Common name	pH tolerance* Acid	Alkaline	Alkali	Source[†]
		Shrubs, cont.			
Acacia redolens	prostrate acacia		x		EBMUD 1990; Svihra and Coate 1996
Aloe spp.	aloe		x		Hoyt 1978
Aloysia triphylla	lemon verbena			x	Hoyt 1978
Arbutus unedo	strawberry tree	x			Hoyt 1978
Artemesia californica	California sagebrush		x		Perry 1989
Atriplex canescens	four-wing saltbush		x		Svihra and Coate 1996
Atriplex lentiflormis spp. *brewerii*	Brewer's saltbush		x	x	Hoyt 1978; Svihra and Coate 1996
Atriplex semibaccata	Australian saltbush		x		Svihra and Coate 1996
Azalea spp.	azalea	x			Hoyt 1978
Azara microphylla	boxleaf azara	x			Hoyt 1978
Baccharis pilularis 'Twin Peaks'	dwarf coyote bush		x		Svihra and Coate 1996
Baccharis spp.	coyote bush		x		Perry 1989
Berberis × *mentorensis*	mentor barberry		x		Svihra and Coate 1996
Berberis spp.	barberry		x		Hoyt 1978
Brunfelsia pauciflora	yesterday-today-and-tomorrow	x			Hoyt 1978
Buxus microphylla var. *japonica*	Japanese boxwood		x		Hoyt 1978; EBMUD 1990; Svihra and Coate 1996
Caesalpinia gilliesii	yellow bird of paradise		x		Hoyt 1978
Caesalpinia pulcherrima	red bird of paradise			x	Hoyt 1978
Callistemon citrinus	lemon bottlebrush		x		EBMUD 1990
Callistemon spp.	bottlebrush		x		Hoyt 1978; Perry 1989
Camellia spp.	camellia	x			Hoyt 1978
Ceanothus griseus	wild lilac			x	Hoyt 1978
Cercocarpus lediflorius	curl-leaf mountain mahogany		x		Kuhns and Rupp 2000
Chaenomeles japonica	flowering quince		x		Hoyt 1978; EBMUD 1990
Cistus salviifolius	sageleaf rockrose		x		Svihra and Coate 1996
Cistus skanbergii	rockrose		x		EBMUD 1990
Cistus spp.	rockrose		x		Hoyt 1978; Perry 1989
Coprosma kirkii	creeping coprosma		x		Perry 1989; Svihra and Coate 1996
Cornus mas	cornelian cherry		x		Kuhns and Rupp 2000
Correa 'Ivory Bells'	white Australian fuchsia		x		Svihra and Coate 1996
Cotinus coggyria	smoke tree		x		Kuhns and Rupp 2000
Cotoneaster apiculatus	cranberry clusterberry		x		Svihra and Coate 1996
Cotoneaster congestus 'Likiang'	Likiang cotoneaster		x		Svihra and Coate 1996
Cotoneaster conspicuus	necklace cotoneaster		x		Svihra and Coate 1996
Cotoneaster horizontalis	rock cotoneaster		x		Svihra and Coate 1996
Cotoneaster lacteus	Parney cotoneaster		x		Svihra and Coate 1996
Cotoneaster microphyllus	rockspray cotoneaster		x		Svihra and Coate 1996
Cytisus canariensis	Canary Island broom			x	Hoyt 1978
Daphne odora	winter daphne	x			Hoyt 1978
Dodonaea viscosa	hopseed bush			x	Hoyt 1978
Elaeagnus angustifolia	Russian olive		x		Kuhns and Rupp 2000
Elaeagnus pungens	silverberry		x	x	Hoyt 1978; EBMUD 1990
Elaeagnus pungens fruitlandii	Fruitland silverberry		x		Svihra and Coate 1996
Escallonia dwarf cvs.	dwarf escallonia		x		EBMUD 1990

Scientific name	Common name	pH tolerance* Acid	Alkaline	Alkali	Source†
		Shrubs, cont.			
Escallonia × *exoniensis* 'Frades'	Frades escallonia		x		EBMUD 1990
Eucalyptus torquata	coral gum			x	Hoyt 1978; Perry 1989
Euonymus fortunei 'Coloratus'	wintercreeper		x		Svihra and Coate 1996
Euonymus japonicus	evergreen euonymus		x		Hoyt 1978
Eurya spp.	eurya	x			Hoyt 1978
Fallugia paradoxa	Apache plume		x		Hoyt 1978
Feijoa sellowiana	pineapple guava		x		Hoyt 1978; Perry 1989; Svihra and Coate 1996
Forsythia × *intermedia*	forsythia		x		EBMUD 1990
Fouquieria splendens	ocotillo		x		Hoyt 1978
Fuchsia spp.	fuchsia	x			Hoyt 1978
Gardenia thunbergia	gardenia		x		Hoyt 1978
Gardenia spp.	gardenia	x			Hoyt 1978
Grevillea × 'Canberra'	grevillea		x		Svihra and Coate 1996
Halimium spp.	halimium		x		Hoyt 1978
Hebe spp.	veronica		x		Hoyt 1978
Hibiscus rosa-sinensis	Chinese hibiscus			x	Hoyt 1978
Hydrangea macrophylla	bigleaf hydrangea	x			Hoyt 1978
Ilex aquifolium	English holly		x		EBMUD 1990
Ilex vomitoria	yaupon		x		Hoyt 1978; EBMUD 1990
Illicium floridanum	anise tree	x			Hoyt 1978
Ixora chinensis	ixora	x			Hoyt 1978
Juniperus californica	California juniper			x	Hoyt 1978
Juniperus chinensis	spreading juniper		x		Kuhns and Rupp 2000
Juniperus chinensis 'Parsonii'	prostrate juniper		x		Svihra and Coate 1996
Juniperus chinensis 'San Jose'	San Jose juniper		x		Svihra and Coate 1996
Juniperus spp.	juniper		x		Hoyt 1978; Perry 1989
Justicia brandegeana	shrimp plant	x			Hoyt 1978
Lagerstroemia indica	crape myrtle			x	Hoyt 1978
Lantana spp.	lantana		x	x	Hoyt 1978; Perry 1989
Laurus nobilis	sweet bay		x		Hoyt 1978; Perry 1989
Lavandula officinalis	English lavender		x	x	Hoyt 1978; Perry 1989
Leptospermum laevigatum	Australian tea tree		x	x	Hoyt 1978; Perry 1989; EBMUD 1990
Leucophyllum texanum	Texas ranger		x	x	Hoyt 1978
Leucothoe axillaris	coast leucothoe	x			Hoyt 1978
Ligustrum spp.	privet		x		Hoyt 1978
Magnolia stellata	star magnolia	x			Hoyt 1978
Mahonia aquifolium	Oregon grape		x		Perry 1989
Mahonia pinnata	California holly grape		x		Svihra and Coate 1996
Mahonia repens	creeping mahonia		x		Svihra and Coate 1996
Melaleuca spp.	melaleuca		x	x	Hoyt 1978; Perry 1989
Myoporum laetum	myoporum			x	Hoyt 1978
Myrica californica	Pacific wax myrtle	x			Hoyt 1978
Myrsine africana	African boxwood		x	x	Hoyt 1978; EBMUD 1990; Svihra and Coate 1996
Myrtus communis	myrtle		x	x	Hoyt 1978; Svihra and Coate 1996
Myrtus communis 'Compacta'	compact myrtle		x		Svihra and Coate 1996

TABLE 5.15. pH tolerance of selected landscape plants, cont.

Scientific name	Common name	pH tolerance*			Source†
		Acid	Alkaline	Alkali	
		Shrubs, cont.			
Nandina domestica 'Nana purpurea'	dwarf purple heavenly bamboo		x		Svihra and Coate 1996
Nerium oleander	oleander		x		Hoyt 1978; Perry 1989
Ochna serrulata	bird's-eye bush	x			Hoyt 1978
Osmanthus fragrans	sweet olive	x	x		Hoyt 1978; EBMUD 1990
Phormium tenax	New Zealand flax		x		EBMUD 1990
Phyllostachys aurea	golden bamboo		x		EBMUD 1990
Pittosporum crassifolium	evergreen pittosporum			x	Hoyt 1978
Pittosporum spp.	pittosporum		x		Perry 1989
Plumbago auriculata	Cape plumbago		x		Perry 1989
Prosopis spp.	mesquite			x	Hoyt 1978
Prunus caroliniana	Carolina laurel cherry		x		Svihra and Coate 1996
Prunus ilicifolia	hollyleaf cherry		x	x	Hoyt 1978; Perry 1989
Prunus lyoni	Catalina cherry		x		Hoyt 1978
Punica granatum	pomegranate		x	x	Hoyt 1978; Perry 1989; EBMUD 1990
Pyracantha coccinea	firethorn			x	Hoyt 1978
Pyracantha spp.	firethorn		x		Hoyt 1978
Rhaphiolepis umbellata	Yeddo hawthorn		x		EBMUD 1990
Rhus integrifolia	lemonade berry		x		Perry 1989; EBMUD 1990
Rhus spp.	sumac		x		Hoyt 1978
Ruscus aculeatus	butcher's broom		x		Svihra and Coate 1996
Salvia greggii	autumn sage		x		Perry 1989
Salvia leucantha	Mexican bush sage		x		Perry 1989
Sollya heterophylla	Australian bluebell creeper		x		Hoyt 1978
Sophora secundiflora	mescal bean		x		Hoyt 1978
Spartium junceum	Spanish broom		x	x	Hoyt 1978
Syringa × *persica*	Persian lilac		x		Hoyt 1978
Tamarix aphylla	athel tree		x		Svihra and Coate 1996
Tamarix parviflora	pink tamarisk		x		Svihra and Coate 1996
Ternstroemia japonica	Japanese ternstroemia	x			Hoyt 1978
Thevetia peruviana	yellow oleander		x		Hoyt 1978
Tibouchina urvilleana	princess flower	x			Hoyt 1978
Ugni molinae	Chilean guava	x			Hoyt 1978
Viburnum spp.	viburnum		x		Hoyt 1978
Viburnum tinus	laurustinus		x	x	Hoyt 1978; EBMUD 1990
Xylosma congestum	shiny xylosma		x		Svihra and Coate 1996
		Vines			
Ampelopsis brevipedunculata	porcelain berry	x			Hoyt 1978
Antigonon leptopus	coral vine			x	Hoyt 1978
Campsis × *tagliabuana* 'Madame Galen'	trumpet vine		x		Svihra and Coate 1996
Cissus antarctica	kangaroo treevine		x		EBMUD 1990
Clematis spp.	clematis		x		Hoyt 1978
Euonymus fortunei radicans	common winter creeper		x		EBMUD 1990; Svihra and Coate 1996
Fallopia baldschuanica	lace vine	x			Hoyt 1978

Scientific name	Common name	pH tolerance* Acid	Alkaline	Alkali	Source†
		Vines, cont.			
Gelsemium sempervirens	Carolina jasmine	x			Hoyt 1978
Hardenbergia spp.	lilac vine		x		Hoyt 1978
Hedera canariensis	Algerian ivy		x		Svihra and Coate 1996
Hedera helix	English ivy		x		Hoyt 1978
Humulus lupulus	hops vine		x		Svihra and Coate 1996
Lonicera japonica 'Halliana'	Hall's honeysuckle		x		Hoyt 1978; Svihra and Coate 1996
Lonicera sempervirens	trumpet honeysuckle	x			Hoyt 1978
Macfadyena unguis-cati	yellow cats-claw creeper		x		Svihra and Coate 1996
Mandevilla laxa	Chilean jasmine		x		Hoyt 1978
Muehlenbeckia complexa	mattress vine			x	Hoyt 1978
Parthenocissus tricuspidata	Boston ivy		x		EBMUD 1990
Philadelphus mexicanus	evergreen mock orange	x			Hoyt 1978
Rosa banksiae	Lady Banks' rose		x		EBMUD 1990
Tropaeolum peregrinum	canary bird flower	x			Hoyt 1978
Vitus vinifera	grape		x		EBMUD 1990
		Herbaceous Plants			
Acanthus mollis	bear's breech	x	x		Hoyt 1978; EBMUD 1990
Achillea millefolium	common yarrow		x		EBMUD 1990
Agapanthus africanus	lily-of-the-Nile		x		Svihra and Coate 1996
Allium schoenoprasum	chives	x			Hoyt 1978
Aloe spp.	aloe			x	Hoyt 1978
Anemone × hybrida	Japanese anemone		x		Hoyt 1978
Anemone coronaria	poppy-flowered anemone	x			Hoyt 1978
Aquilegia chrysantha	golden columbine	x			Hoyt 1978
Arctotis hybrids	African daisy		x		EBMUD 1990
Arenaria balearica	Corsican sandwort		x		Hoyt 1978
Arundo donax	giant reed			x	Hoyt 1978
Aspidistra elatior	cast-iron plant		x		EBMUD 1990
Aster amellus	Italian aster		x	x	Hoyt 1978
Begonia spp.	begonia	x			Hoyt 1978
Bergenia crassifolia	winter-blooming bergenia		x		EBMUD 1990
Bergenia spp.	bergenia	x			Hoyt 1978
Calendula officinalis	calendula		x	x	Hoyt 1978
Callistephus chinensis	China aster		x		Hoyt 1978
Centaurea spp.	centaurea		x		Hoyt 1978
Ceratostigma plumbaginoides	dwarf plumbago		x		EBMUD 1990
Chrysanthemum frutescens	marguerite			x	Hoyt 1978
Chrysanthemum maximum	Shasta daisy			x	Hoyt 1978
Chrysanthemum parthenium	feverfew		x		EBMUD 1990
Chrysanthemum spp.	chrysanthemum		x		Hoyt 1978
Coleus blumei	coleus	x			Hoyt 1978
Coronilla glauca	crown vetch		x		Hoyt 1978
Cortaderia selloana	pampas grass			x	Hoyt 1978
Cosmos bipinnatus	cosmos			x	Hoyt 1978
Cymbalaria muralis	Kenilworth ivy		x		Hoyt 1978
Cynodon dactylon	bermudagrass			x	Hoyt 1978

TABLE 5.15. pH tolerance of selected landscape plants, cont.

Scientific name	Common name	pH tolerance* Acid	Alkaline	Alkali	Source[†]
		Herbaceous Plants, cont.			
Delosperma alba	white trailing ice plant		x		EBMUD 1990
Delphinium nudicaule	scarlet larkspur	x			Hoyt 1978
Dianthus spp.	carnation		x		Hoyt 1978
Dicentra spectabilis	bleeding heart	x			Hoyt 1978
Dietes vegeta	fortnight lily		x		EBMUD 1990
Dimorphotheca spp.	African daisy			x	Hoyt 1978
Dodecatheon clevelandii	shooting star	x			Hoyt 1978
Drosanthemum floribundum	rosea ice plant		x		EBMUD 1990
Echium fastuosum	pride of Madeira		x		Svihra and Coate 1996; Perry 1989
Eriogonum fasciculatum	California buckwheat		x	x	Hoyt 1978
Eschscholzia californica	California poppy		x		EBMUD 1990
Euphorbia characias wulfenii	spurge		x		EBMUD 1990
Euyrops pectinatus	yellow bush daisy		x		Svihra and Coate 1996
Festuca californica	California fescue		x		EBMUD 1990
Francoa ramosa	maiden's wreath	x			Hoyt 1978
Gazania spp.	gazania			x	Hoyt 1978
Geranium incanum	cranesbill		x		EBMUD 1990
Gerbera jamesonii	Transvaal daisy			x	Hoyt 1978
Gladiolus hortulanus	gladiolus	x			Hoyt 1978
Gypsophila elegans	showy babysbreath		x		Hoyt 1978
Helleborus lividus	hellebore	x			Hoyt 1978
Hemerocallis spp.	daylily		x		EBMUD 1990
Hesperaloe parviflora	red yucca			x	Hoyt 1978
Heterocentron roseum	Spanish shawl	x			Hoyt 1978
Heuchera sanguinea	coral bells		x		Hoyt 1978
Hunnemannia fumariifolia	Mexican tulip poppy		x		Hoyt 1978
Hypericum calycinum	creeping St. Johnswort		x		Svihra and Coate 1996
Hypericum spp.	St. Johnswort		x		Perry 1989; EBMUD 1990
Iris foetidissima	Gladwin iris		x		EBMUD 1990
Iris hybrids	iris		x		EBMUD 1990
Iris japonica	Japanese iris		x		Hoyt 1978
Kniphofia uvaria	red-hot poker		x		Svihra and Coate 1996
Lampranthus aurantiacus	ice plant		x		EBMUD 1990
Lampranthus spectabilis	trailing ice plant		x		EBMUD 1990
Lathyrus odoratus	sweet pea		x		Hoyt 1978
Leucojum autumnale	snowflake	x			Hoyt 1978
Lilium candidum	madonna lily		x		Hoyt 1978
Lilium henryi	Henry's lily		x		Hoyt 1978
Lilium longiflorum	Easter lily	x			Hoyt 1978
Lilium regale	regal lily		x	x	Hoyt 1978
Limonium perezii	sea lavender		x		EBMUD 1990
Lobularia maritima	sweet alyssum		x		EBMUD 1990
Lobularia spp.	lobularia		x		Perry 1989
Lotus corniculatus	bird's foot trefoil	x			Hoyt 1978
Malephora crocea	ice plant		x		EBMUD 1990

Scientific name	Common name	pH tolerance*			Source†
		Acid	Alkaline	Alkali	
Herbaceous Plants, cont.					
Mesembryanthemum spp.	ice plant			x	Hoyt 1978
Mirabilis jalapa	four o'clock		x		EBMUD 1990
Miscanthus sinensis	Japanese silver grass	x			Hoyt 1978
Myosotis sylvatica	forget-me-not		x		EBMUD 1990
Nerine spp.	nerine	x			Hoyt 1978
Nicotiana alata	flowering tobacco		x		Hoyt 1978
Oenothera spp.	evening primrose		x	x	Hoyt 1978
Pelargonium spp.	geranium		x		Hoyt 1978
Pellaea andromedaefolia	coffee fern		x		Hoyt 1978
Pennisetum alopecuroides	fountain grass		x		EBMUD 1990
Pennisetum spp.	fountain grass		x		Perry 1989
Phlomis fruiticosa	Jerusalem sage		x		Svihra and Coate 1996
Phormium tenax	New Zealand fax			x	Hoyt 1978
Phyla nodiflora	lippia		x		EBMUD 1990
Portulaca grandiflora	rose moss			x	Hoyt 1978
Romneya coulteri	matilija poppy		x		EBMUD 1990
Rudbeckia hirta	gloriosa daisy		x		EBMUD 1990
Salpiglossis sinuata	painted tongue		x		Hoyt 1978
Santolina chamaecyparissus	lavender cotton		x		EBMUD 1990; Svihra and Coate 1996
Santolina virens	green lavender cotton		x		EBMUD 1990; Svihra and Coate 1996
Scabiosa spp.	pincushion flower		x		Hoyt 1978
Scaevola 'Mauve Clusters'	fan flower		x		EBMUD 1990
Scilla peruviana	Peruvian scilla		x		EBMUD 1990
Sedum spathulifolium	stonecrop		x		EBMUD 1990
Sempervivum spp.	houseleek		x		Hoyt 1978
Senecio cineraria	dusty miller		x		Hoyt 1978
Teucrium fruticans	bush germander		x		EBMUD 1990
Thalictrum dipterocarpum	Chinese meadow rue	x			Hoyt 1978
Thymus spp.	thyme		x		Hoyt 1978
Tropaeolum spp.	nasturtium		x		Hoyt 1978
Tulbaghia violacea	society garlic		x		Hoyt 1978
Verbena tenuisecta	moss verbena		x		EBMUD 1990
Verbena spp.	verbena		x	x	Hoyt 1978
Vinca spp.	periwinkle		x		Hoyt 1978
Viola spp.	violet	x			Hoyt 1978
Zantedeschia spp.	calla	x			Hoyt 1978
Zauschneria californica	California fuschia		x		EBMUD 1990
Zephyranthes candida	zephyr flower		x		Hoyt 1978
Zephyranthes spp.	zephyr flower	x			Hoyt 1978
Zinnia elegans	zinnia		x		Hoyt 1978

Notes:

* Acid = pH 5–6.5; alkaline = pH 7.5–8.2; alkali = pH > 8.3, SAR > 6.

† For Svihra and Coate 1996, only trees rated 4 and 5 for alkaline tolerance are included; for Kuhns and Rupp 2000, only plants rated as high alkaline tolerance are included.

TABLE 5.16. Summary of pH-related problems

Soil pH	Symptoms	Diagnosis	Occurrence/ aggravating factors	Look-alike disorders	Treatment
			Acid soils		
<5.0	Typically expressed in micronutrient toxicities in sensitive plants.	Test foliage for microelement concentration.			Incorporate lime into soil to raise pH to desired level.
	Mottled chlorosis of leaves, followed by necrotic spots.	Manganese toxicity.	Wet soils.	Iron deficiency.	Incorporate lime into soil to raise pH above 5.5. Improve drainage.
	Roots are short, thick and stubby.	Aluminum toxicity.	pH <4.5.	Surflan (pre-emergent herbicide) use and Ca deficiency cause similar effects on roots. Foliage may appear to be P deficient because of inability to take up P.	Incorporate lime into soil to raise pH above 5.5. Apply phosphorus.
	Interveinal chlorosis on foliage. Stunting, reduced branching; thickening and dark coloration of roots.	Copper toxicity.	Heavy spraying of Bordeaux fungicides or copper sulfate fertilization.	Iron deficiency.	Incorporate lime into soil to raise pH above 5.5.
			Alkaline soils		
>7.5	Typically expressed in micronutient deficiencies in sensitive plants.	Test foliage for micronutrient concentration.			In noncalcerous soils, incorporate sulfur to lower pH over time.
	Interveinal chlorosis, most severe on new growth.	Iron deficiency.	Cold soil tempertures.	Similar symptoms can be caused by root disease or may be normal in early spring.	See the section "Nutrient Disorders"
	New leaves yellow, with wide green bands along veins.	Manganese deficiency.		Similar symptoms can be caused by root disease or may be normal in early spring.	See the section "Nutrient Disorders"
	Leaves chlorotic, sometimes mottled with necrotic spots; leaves small; shortened internodes.	Zinc deficiency.		Similar symptoms can be caused by root disease or may be normal in early spring	See the section "Nutrient Disorders"

Low Temperature (Chilling Injury, Frost Injury, Freezing Injury)

Plant cells are injured when temperature declines below a critical level for a species. Injury that occurs at or below the freezing point (32°F, 0°C) is called *frost injury* or *freezing injury*. Injury above the freezing point is called *chilling injury*. In this publication, all three types of injury are referred to as low temperature injury (see the sidebar "Chilling, Frost, and Freezing Injury: What's the Difference?" p. 138).

Since temperature is the least controllable environmental factor in a landscape, it is important to choose species that are cold hardy for the location. Refer to publications such as the *California Master Gardener Handbook* (Pittenger 2002), and the *Sunset Western Garden Book* (Brenzel 2001) for information on plant climate zones and evaluations of species sensitivity or tolerance. The *USDA Plant Hardiness Zone Map* (USDA 1990) identifies climate zones for the United States based on winter minimum temperatures.

Figure 5.62. In some cases, only the most exposed part of a plant canopy is injured following low temperatures. Cold injury on these oleander *(Nerium oleander)* is confined largely to the more exposed upper half of the canopy.

Symptoms

Low temperature injury can occur on all parts of a plant: leaves, shoots, flowers, buds, fruit, bark, wood, and roots. Although broadleaf evergreen species are generally more sensitive to low temperature injury, deciduous and coniferous plants can be injured as well.

The plant parts injured and the degree of injury depends on critical temperatures, species, the duration of low temperature, temperature changes, plant condition (acclimation or hardening), age, hydration, and time of year (fig. 5.62). Tender new leaves exposed to critical low temperatures appear water-soaked and may turn black. Mature leaves turn reddish brown to dark brown or nearly black (fig. 5.63A). Leaves remain attached for a period of time in some species or drop off in others; leaves of deciduous species typically abscise when subjected to cold. The youngest leaves and shoots are most sensitive to injury. Chilling temperatures (above 32°F, or 0°C) cause the foliage of some species to turn red or purple (fig. 5.63B).

In spring, when plants break dormancy, the emerging shoots are highly susceptible to low temperature injury. Often the entire shoot is killed. If stem bark tissues are not killed, new shoots may develop from dormant or epicormic buds (fig. 5.64).

Flowers buds are less hardy than vegetative buds and may be injured at temperatures higher than those that injure shoots. Injury may occur either during winter dormancy or during the spring blooming period. Plants fail to bloom or blossoms are fewer and fruit set is reduced. Conversely, if certain temperate plants receive insufficient chilling periods (temperatures below 45°F, or 7°C), growth and flowering may be delayed or reduced in spring.

Both wood and bark tissue can be damaged by low temperatures. Longitudinal splitting of wood (frost cracks), black heart of stems (xylem darkening), separation of wood along annual rings (cup shakes), and bark splitting have resulted from low temperatures.

Although roots may be injured, low temperature injury to roots is not common. Typically, the soil provides sufficient insulation to moderate temperatures in the root zone. The roots of plants in containers and raised beds are susceptible to freezing damage, however.

Occurrence

Low temperature injury results when tissues are chilled below the temperature they are adapted to tolerate. Damage is both physiological and anatomical: ice crystals form within cells, causing dehydration and disruption of membranes and organelles. Although plants can be injured at any time of the year, the most critical periods are

Figure 5.63. Low temperatures can cause the foliage of sensitive species to turn from green to red-brown in a short time. Here, the entire canopy of Australian brush cherry *(Syzygium paniculatum)* has turned red-brown following several nights of critically low temperatures (A). In princess flower *(Tibouchina urvilleana)*, leaves turn red at temperatures above freezing (B).

Figure 5.64. Although leaves and shoots may be killed by low temperatures, trunk tissues may survive. Here, low temperature caused extensive dieback (A) in the canopy of myoporum *(Myoporum laetum)*, but after several months epicormic shoots developed along the trunk (B).

Figure 5.65. Low-lying locations are subject to lower temperatures than adjacent higher areas. The Strybing Arboretum nursery is located in one of the lowest-lying areas of Golden Gate Park and is considered to be one of the coldest locations in San Francisco.

spring and autumn, the coldest part of winter, and when temperatures decline rapidly after a warm period.

Cold temperatures in the fall are a problem if plants have not hardened, or become acclimated. Unacclimated plants may be injured at temperatures well above those tolerated by hardened plants. Conditions favoring early cessation of growth (low soil nitrogen levels and low soil moisture content) also favor early development of cold hardiness.

In spring, sensitive new growth may be injured during late frosts. Typically, injury occurs when a warm period is followed by cold temperatures.

Wood and bark can develop frost cracks when the air temperature drops substantially during the dormant period. The inner wood remains relatively warm while the outer wood becomes cold and contracts rapidly, causing cracks or splits in the trunk (see Allen, Morrison, and Wallis 1996). Callus tissue that forms after the frost crack appears as raised, black lines on the stem.

The onset of low temperature injury may be short (overnight) or of greater duration (several days). Low-lying locations are likely to have lower temperatures than surrounding higher areas (fig. 5.65). Young plants are more sensitive to injury than mature specimens, and plants that are isolated have a higher potential for injury than those in groups (figs. 5.66–67).

The roots of plants growing in the ground are generally protected from freezing. Plants in containers or raised beds, however, have a higher potential to sustain root zone injury than those in the field.

Look-Alike Disorders

The symptoms of water deficit, gas injury, chemical injury, root disease, anthracnose, and mechanical injury to roots are similar to the symptoms of low temperature injury.

Diagnosis

- Inspect injured plants carefully and determine which parts are damaged and the extent of injury. Are symptoms consistent with those of low temperature injury?

Figure 5.66. Trees occurring alone have a higher potential of being injured by low temperatures than trees in groups or next to protective structures. This New Zealand Christmas tree *(Metrosideros excelsus)* may not have sustained the same level of injury if it was in a group of trees or next to a building.

Figure 5.67. Young trees are particularly sensitive to low temperature injury. These young Indian laurel fig *(Ficus microcarpa)* were severely damaged by cold temperatures in San Jose, California.

- Investigate the low temperature tolerance of the species. Many plant manuals contain this information, including the *California Master Gardener Handbook* (Pittenger 2002) and the *Sunset Western Garden Book* (Brenzel 2001).
- Determine whether temperatures exceeding the tolerance of the species have occurred in the area. Make sure you know where the temperature measurement was taken and determine whether it is equivalent to the site in question.
- Determine whether symptom onset followed immediately or some time after a critical low temperature or period of low temperatures. Be aware that some injuries result from repeated episodes of low temperature exposure.
- Inspect bark and bud tissues. Do they appear water-soaked or dehydrated? Are they dark brown or black? Even though the foliage may be dead, bark and bud tissues may not have been killed, and regrowth may follow.
- For palms, be sure to inspect the apical bud. Even though many fronds can be killed, regrowth may occur if the apical bud is still alive. Conversely, if the apical bud has been killed, recovery of that stem is not possible.

Sensitive and Tolerant Species

Some species tolerate temperatures well below freezing, while others are injured at temperatures close to freezing (fig. 5.68). Many plant manuals provide information regarding plant hardiness. When diagnosing low temperature injury or when selecting plants for a landscape, be sure you know the critical low temperature for the species.

Remedies

After low temperature damage has occurred, not much can be done to save injured plant parts. In some cases, only foliage or tender shoots will be injured. In other cases, the whole plant will be killed.

Do not prune parts or remove the plant until you are sure of the extent of injury. Many species can produce new foliage after canopy dieback (fig. 5.69). Use a sharp

Figure 5.68. Species vary in tolerance to cold temperatures. Here, Canary Island pine *(Pinus canariensis)* (left) was injured, but Monterey pine *(Pinus radiata)* was not (A). After 6 months, however, the Canary Island pine recovered (B).

Figure 5.69. After sustaining considerable foliar injury from cold winter winds (A), these coast redwood *(Sequoia sempervirens)* recovered in the spring (B). It is important to postpone any pruning or plant removals until sufficient time is allowed for recovery.

Figure 5.70. Although sprinkler irrigation can be used to protect nursery plants from low temperature injury, this method is not practical in most landscapes.

knife to check wood tissues below the bark. If they are dry and brown, they are likely dead. If they are hydrated and white or light brown, then they are still alive. Even if the whole plant appears to be dead, it is prudent to wait until the spring and then look for signs of regrowth. This is particularly important for palms. If no regrowth is noted, prune out dead parts or remove the plant. Replant with tolerant species.

In some cases, further damage may be avoided by providing protection around the plant, particularly if radiation frost is expected (see the sidebar "Chilling, Frost, and Freezing Injury: What's the Difference?"). Irrigating the soil thoroughly, circulating air using a fan, or providing coverings to retain heat may prove beneficial. An increase in temperature of 2° to 4°F (1° to 2∞ C) can make a substantial difference. Well-hydrated plants are generally more tolerant of low temperature injury than water-stressed plants, and moist soils have a higher heat capacity that dry soils. Avoid stimulating new growth in late summer or early fall by fertilizing or pruning. For container plants, consider moving the plant to a warmer location, such as next to a wall or under a roof line. Wind machines, sprinkler irrigation, and orchard heaters are used to protect plants in nurseries and orchards, but they are typically not practical in urban landscapes (fig. 5.70).

Chilling, Frost, and Freezing Injury: What's the Difference?

Chilling Injury

Chilling injury is damage to plant parts caused by temperatures above the freezing point (32°F, 0°C). Plants of tropical or subtropical origin are most susceptible, such as avocado (*Persea* spp.), banana (*Musa* spp.), mango *(Mangifera indica)*, African violet (*Saintpaulia* spp.), okra *(Abelmoschus esculentus)*, and tomato *(Lycopersicon esculentum)*. Chilling-injured leaves may become purple or reddish and in some cases wilt. Both flowers and fruit of sensitive species can be injured.

Frost and Freeze Injury

Frost injury and freeze injury are closely related. Frost damage occurs during a radiation freeze; freeze damage occurs during an advection freeze. In both cases, ice crystals form in plant tissues, dehydrating cells and disrupting membranes.

Advective freezes occur when an air mass whose temperature is below freezing moves into an area and displaces warmer air, causing the temperatures of plants to become low enough for ice crystals to form within their tissues.

Radiation freezes occur on clear, calm nights when plants radiate (lose) more heat into the atmosphere than they receive. This creates a temperature inversion in which cold air close to the ground is trapped by warmer air above it (the temperature of the air increases with altitude). When the air temperature at plant level is near or below freezing, the temperature of the plants is likely to be colder than the temperature of the air. If plants become sufficiently cold, the water in them freezes and cells are damaged. The frost that appears on plants is simply ice crystals that form on the plant surface, the equivalent of dew forming at temperatures above freezing. The frost itself does not damage plants; plants are damaged by ice crystals that form within their tissues.

SUNBURN

Sunburn is injury to aboveground plant parts (leaves, bark, flowers, and fruit) caused by excessive exposure to solar radiation. Injury results when tissues become dehydrated after being heated beyond a critical limit. Injury from heat sources, such as steam, fire, or paving equipment, is considered to be thermal injury (see "Thermal Injury," p. 147).

Symptoms

Sunburn may cause leaf discoloration and necrosis. The epidermis may appear glazed, turning a silvery or reddish brown color (fig. 5.71). In advanced cases, distinctive necrotic areas develop on the leaf blade (fig. 5.72). Marginal necrosis may develop, usually starting at the leaf tip, where transpiration is highest, and progressing along the entire margin. The onset of symptoms is usually rapid.

Sunburned bark initially appears discolored (often reddish brown) and then becomes dry. Cracking and peeling is typical, and damage is usually most severe on the south or southwest sides of branches and trunks (fig. 5.73). Damage from wood-boring insects and wood decay often appears on sunburned bark (fig. 5.74).

Sunburn on flowers and fruit appears as water-soaked areas on the most exposed surfaces (fig. 5.75). Eventually, damaged

Figure 5.71. Sunburn can cause reddening of the most sun-exposed parts of leaves. Reddening of the leaves of this newly planted winged euonymous *(Euonymous alata)* was attributed to sunburn.

Figure 5.72. Sunburn can cause red-brown necrotic areas on leaves of sensitive species. Injury to these rhododendron (*Rhododendron* sp.) leaves resulted from afternoon sun exposure during days when air temperatures rose above 90°F (32°C).

Figure 5.73. Sunburn causes drying, cracking, and checking of bark. Many of the ash (*Fraxinus* sp.) in this parking lot were sunburned on the southwest side of the trunk (A). Sunburned bark is often killed, and injury may appear cankerlike (B).

Figure 5.74. Infestation by wood-boring insects and wood decay often follow sunburn injury to bark.

flower tissue turns brown and shrivels, while fruit tissue often appears rotten.

Occurrence

Although most common in summer, sunburn can occur at any time of the year, even in winter. It typically occurs on plant parts receiving greatest exposure to the sun, usually on the south and southwest sides.

Sunburn injury is most severe during periods of high temperature. Plants that tolerate full sun conditions may be damaged when temperatures reach a critical level. Tree trunks in closely-spaced container nursery rows are frequently shaded, and when trees are removed, sun-exposed bark is prone to sunburn. Trees that have been excessively pruned or experienced root injury are also prone to sunburn injury.

During cool, spring conditions, some species may develop thin leaves that are sensitive to sunburn (see Hagen 2000). When a period of high temperature follows, leaves and shoots directly exposed to the sun may become sunburned (fig. 5.76). Species reported to be injured in this manner include coast redwood (*Sequoia sempervirens*), Douglas fir (*Pseudotsuga menziesii*), white fir (*Abies concolor*), maples (*Acer* spp.), and horsechestnut (*Aesculus hippocastanum*).

Figure 5.75. Sunburn injury can occur on flowers and fruit. The petals of this rose (*Rosa* sp.) were damaged when a hot day (>90°F, 32°C) followed cool days (<60°F, 15°C).

Deciduous trees may sustain sunburn injury during the winter months. When branches previously protected by leaves are exposed to direct sun, bark tissues may be sunburned. Similarly, when branches or trees shading other branches or trees are removed, sunburn may result. This can be particularly severe on young trees, thin-barked species, and water-stressed trees.

Look-Alike Disorders

Disorders that resemble sunburn injury include water stress, salt stress, specific ion toxicity, gas injury, and herbicide injury. Sunburn injury to bark is very similar to that caused by sunscald (see "Sunscald," p. 144).

Figure 5.76. New growth is particularly sensitive to sunburn in certain species. The young shoots on the southwest side of this coast redwood (*Sequoia sempervirens*) were severely sunburned during a heat wave (>90°F, 32°C) in late June (A). Close-up of injured shoot (B).

Sunburn, Sunscald, High Temperature Injury, and High Light Injury: What's the Difference?

In this publication, sunburn, sunscald, high temperature injury, and high light injury are defined based on how they have been described and used in other publications and based on our assessment of their application in diagnosing abiotic disorders.

Sunburn

Solar radiation exposure can generate critically high temperatures in plant tissues, leading to dehydration and death. Sunburn is injury to aboveground plant parts (leaves, bark, flowers, and fruit) caused by excessive exposure to solar radiation. High ambient temperatures are closely linked to sunburn injury. Plant water deficits increase the potential for injury, but sunburn can occur on sensitive species when soil moisture levels are adequate. Unlike sunscald, sunburn is not preceded or followed by freezing temperatures.

Sunscald

Sunscald is winter injury of the bark of limbs and trunks of woody plants, principally in northern latitudes. Rapid changes in temperature are believed to cause death of cambial cells, leading to a separation of bark and wood. Tissues are injured when freezing temperatures precede or follow daytime warming. Factors thought to contribute to sunscald include thin bark, trunk injuries (wounds), root injury, borers, and plant water deficits. Of these, water deficits have been found to be a crucial factor contributing to injury in newly planted trees (Roppolo and Miller 2001). Sunscald injury is most common on the south and southwest sides of trunks and branches.

High Temperature Injury

High temperature injury, or thermal injury, is caused by critically high ambient temperatures. Leaves, flowers, fruit, bark, wood, and roots can be damaged. When a critical temperature is reached or exceeded for a species, cells experience chemical alterations, membrane disruption, dehydration, and death. High ambient temperatures can be generated from sources such as solar radiation, heat releases from vents, pipes, or other equipment, and also by the decomposition of organic matter. In this publication, high temperature injury to aboveground plant parts caused by direct solar radiation exposure is considered to be sunburn injury. Injury caused by other heat sources is considered to be thermal injury.

High Light Injury

High light injury is foliar chlorosis caused by high light intensity. In sensitive species, chlorophyll is photooxidized in epidermal and upper palisade cells as light intensity increases above a critical level (Treshow 1970). Typically, high light injury occurs when shade-requiring plants are placed in the sun. Injury may occur independent of ambient and foliar temperatures or plant moisture status. The critical level for injury varies with species, acclimatization, and maturity of foliage.

Diagnosis

- Identify the species and determine whether it is reported to be sensitive to sunburn injury.
- Investigate recent weather patterns and determine whether critical high temperatures preceded the onset of symptoms. Assess whether a period of cool temperature was followed by a period of much higher temperature.
- Which tissues are injured? Are the most exposed parts of the plant injured? Are shaded parts of the plant injured?
- Evaluate the soil moisture status. Is the soil dry? Does the plant look water-stressed? Did a windy period occur prior to injury?
- Was the plant recently pruned? Was an adjacent tree (or shading structure) removed? Did injury occur after leaf drop (for deciduous species)? Are reflective surfaces contributing to exposure level (fig. 5.77)?

Sensitive and Tolerant Species

Species vary in their tolerance to sunburn injury: excessive exposure for one species may be innocuous for another. Tolerance also varies with stage of plant development and previous environmental history. Plants that tolerate full sun conditions may be sunburned when critically high temperatures occur. Refer to plant manuals for information regarding the sensitivity of particular species to sunburn.

Species planted at the fringes of their planting zones or in less than optimal growing conditions are susceptible. For example, Grecian laurel (*Laurus nobilis*), a thin barked-tree, is well-adapted to a Mediterranean climate, but when planted as a street tree in small planting pits with south, west, or reflective exposures, it frequently sustains sunburn injury. Camphor (*Cinnamomium camphora*) and ash (*Fraxinus* spp.) have been observed to be sensitive to sunburn injury on bark.

Trees that have been heavily pruned or pollarded are prone to sunburn injury. Olive and many conifer species are particularly susceptible.

Remedies

If possible, reduce temperatures in the plant environment. Add shade, improve air circulation, and increase humidity. If soil moisture levels are low, irrigate. Protect sensitive plants from wind (if practical). Select species for tolerance to sunburn. Bark may be protected with whitewash (for additional remedies, see "Sunscald," p. 144).

Figure 5.77. Planting sensitive species next to reflective surfaces increases the potential for sunburn injury.

Sunscald

Figure 5.78. Mild sunscald symptoms include red discoloration of smooth bark.

Sunscald is damage to bark caused by rapid temperature fluctuations during the winter. Bark exposed to freezing temperatures at night can be injured when warmed by the sun to a critical level during the day. Although the mechanism of injury is unclear, it is thought that sensitive cambial cells are killed because they are unable to adjust to rapid temperature changes. Subsequently, bark separates from the underlying wood. It is also reported that injury may occur when freezing temperatures (either at night or as a result of cloud cover) follow bark warming during the day.

Symptoms

Initially, sunscald damage may appear as reddish brown discoloration of the bark (fig. 5.78). The bark shrinks, appears sunken, splits, and then peels back in chunky patches, exposing sapwood under-

Figure 5.79. Bark injury found on the trunk of this Chinese pistache *(Pistacia chinensis)* was attributed to sunscald (A). Sunscalded bark peels back in chunky patches (B). Sunburn and sunscald symptoms are similar.

neath (fig. 5.79). Cankers may develop. In severe cases, the entire trunk may be girdled, or individual branches may die.

Figure 5.80. Sunscald damaged tissue may be further injured by wood-boring insects and decay.

Figure 5.81. Newly planted trees are particularly susceptible to sunscald injury.

Occurrence

Sunscald injury is usually found on the south, southwest, or west side of trunks and branches and is most common on water-stressed trees (Roppolo and Miller 2001). Often, sunscald-damaged tissues are further injured by wood-boring insects and wood decay or canker fungi (fig. 5.80). Since sunscald requires freezing temperatures, injury is most common in northern latitudes and is not likely to be found in mild-winter areas.

Young trees with thin bark are particularly prone to sunscald (fig. 5.81). Newly planted or recently transplanted trees are notably susceptible. Container-grown trees sunscald more often than field-grown trees.

Look-Alike Disorders

Sunscald is very similar to sunburn injury to bark. Mechanical damage on the lower trunk, frost cracks, borers, canker-forming pathogens (*Eutypa*, *Nectria*, and fireblight), water injury from irrigation sprinklers, and rubbing injury from stakes may cause sunscaldlike injury.

Diagnosis

Diagnosis requires knowledge of contributing factors: recent weather conditions, moisture status of soil and plant, recent pruning, planting dates, stock type, and potential for root injury. Evaluate the location of injury on the plant and the potential for direct exposure of bark to the sun. Determine whether other species in the area have been similarly affected.

Sensitive and Tolerant Species

Genera sensitive to sunscald include *Liriodendron*, *Acer*, *Tilia*, *Prunus*, *Pyrus*, *Malus*, *Juglans*, *Ulmus*, *Photinia*, and *Laurus*, particularly when grown as standards.

Nursery-grown trees are less prone to sunscald when they possess the following characteristics:

- well-developed root system free of significant kinks or girdles
- foliage retained along the trunk

Figure 5.82. In some orchards, trunks and limbs are painted white to prevent sunscald or sunburn injury.

- exposure to low temperatures prior to planting (hardened-off)
- adequate soil moisture before and after planting

Remedies

Avoiding sunscald is key. Once sunscald injury has occurred, the tree responds by developing callus tissue. Postinjury treatment may prevent further injury but will not repair damage. To avoid sunscald, take the following steps:

- Plant healthy trees that are well adapted to the climate.
- Avoid excessive removal of lower foliage or branches for a few years after planting.
- Keep trees adequately irrigated. Make sure that the root ball and the field soil surrounding the root ball are moist after planting.
- Protection may be achieved by painting bark with 50-50 mixture of white latex paint and water. Paint newly exposed limbs (fig. 5.82). Wrap trunks with light-colored, reflective, protective material. Avoid dark-colored wraps. Once wraps are removed, the bark may still be sensitive to sunscald. Continue to protect until the bark is hardened off.
- Apply a layer of coarse organic mulch 4 to 6 inches (10 to 15 cm) deep around sensitive trees. This will help retain soil moisture, keep roots cooler, and reduce reflected light and heat. Avoid placing mulch against the trunk or stem.

Thermal Injury

Thermal injury occurs when temperatures in the canopy or root zone rise above a critical level due to fire, steam, or heat released by equipment, vents, or other sources.

Most plant cells are killed at temperatures from 122° to 140°F (about 50° to 60°C), depending on the species, age of plant tissue, duration of high temperature exposure, degree of hydration, and occurrence of other weakening factors such as wood-boring beetles and decay fungi. Damage is usually acute, severe, and typically not difficult to diagnose.

Since fire is the most common cause of thermal injury, it is addressed separately from steam and heat injury.

Figure 5.83. Foliage injured by fire can appear scorched, crisp, and blackened.

Figure 5.84. Fire under this pine (*Pinus* sp.) caused one section of the canopy to turn uniformly brown.

Fire

Symptoms

Fire can injure plants directly by combustion or indirectly by heat release. Symptoms include scorched, crisp, brown foliage and charred or blackened stems and branches (figs. 5.83–84).

Trunk damage may not be seen immediately, but heat-damaged bark will separate from underlying tissue. As the trunk expands and callus develops, bark begins to slough off, exposing wood beneath.

Symptom severity depends on heat intensity, duration of exposure, plant moisture content, and how well the species tolerates fire. Plant water content often determines the depth of injury in wood. If trees or shrubs are water-stressed, the degree of injury is greater. Heat-damaged wood is prone to decay and infestation by wood-boring insects.

In grass fires, trees usually exhibit scorched foliage near the ground and in the lower canopy (fig. 5.85). In crown fires, symptoms are observed throughout the canopy.

Occurrence

All plants are subject to fire injury, although species with a high concentration of volatile hydrocarbons in the leaves tend to ignite more readily if the fire is hot enough or exposure long enough. Young trees with little biomass or thin protective bark are most severely injured.

Landscapes located in or near wildlands or in hot, dry chaparral zones are at high risk to wildfire exposure (fig. 5.86). Flammable mulches such as straw, peat moss, and other dry, compact materials can ignite and damage plants. A progression of low to tall plants creates a fire ladder that promotes crown fires. Low-branched trees and trees with dead branches increase the potential of fire injury.

Figure 5.85. Damage to coast live oak *(Quercus agrifolia)* from a grass fire was confined to the lower part of the canopy.

Figure 5.86. Landscapes in wildland areas are at high risk to fire injury. Trees next to this fire-damaged home in an urban-wildland interface area were injured by fire. Damage is distinctive: foliage closest to the fire is most severely injured.

Look-Alike Disorders

Typically, injury from fire is distinctive and not easily confused with other disorders. However, look-alike disorders include herbicide injury (contact or systemic), acute water stress, cold temperature injury (freeze damage), and severe fireblight infections.

Diagnosis

- Examine the pattern of damage. Is it localized in one area or distributed throughout the canopy? Is more than one species affected?
- Did the symptoms occur recently? Did they appear following a fire? Is there char on the trunk or branches?
- Look for possible sources of fire (fig. 5.87) or signs of fire.
- In cases of heat damage to bark, injury may not be apparent until several months after the fire, and the extent of injury may not be known until the following season.
- The level and size of trunk char can be used to evaluate cambium damage:

  Light char: Spotty charring with scattered pitting of bark.

  Medium char: Continuous charring with minor reduction in bark thickness.

  Heavy char: Continuous charring with pronounced reduction in bark thickness with underlying wood sometimes exposed.

Light or medium char usually indicates that the intensity of fire was not sufficient to seriously injure cambium tissue. Heavy char requires further evaluation by removing a small area of intact bark (fig. 5.88). If the underlying cambium is yellowish rather than white or pink, it is likely that the tissue is dead or seriously injured. In some cases, a fermentation aroma may be detected. If bark has separated from the wood, the cambium is probably dead (fig. 5.89). Species response to fire is often determined by bark thickness and trunk diameter.

TABLE 5.17. Fire fuel volume of selected landscape plants

Scientific name	Common name
Low fuel volume at maturity	
Acer spp.	maple
Arbutus unedo	strawberry tree
Arctostaphylos spp. (low shrub species)	manzanita
Atriplex spp.	saltbush
Betula spp.	birch
Calocedrus decurrens	incense cedar
Carpobrotus edulis, *Lampranthus* spp., *Drosanthemum* spp.	ice plant
Cercis occidentalis	western redbud
Cistus spp.	rockrose
Citrus spp.	citrus
Convolvulus cneorum	bush morning glory
Feijoa sellowiana	pineapple guava
Myoporum spp.	myoporum
Nerium oleander	oleander
Pittosporum spp.	pittosporum
Populus spp.	cottonwood
Populus spp.	poplar
Populus tremuloides	quaking aspen
Punica granatum	pomegranate
Quercus agrifolia	coast live oak
Quercus douglasii	blue oak
Quercus kellogii	California black oak
Quercus lobata	valley oak
Rhamnus alaternus	Italian buckthorn
Ribes spp.	currant, gooseberry
Salix spp.	willow
Schinus molle	California pepper tree
Schinus terebinthifolius	Brazilian pepper tree
Simmondsia chinensis	jojoba
Tecomaria capensis	Cape honeysuckle
Umbellularia californica	California bay laurel
Low-growing species with moderate fuel volume at maturity	
Arctotheca calendula	Cape weed
Baccharis pilularis	prostrate coyote bush
Coprosma kirkii	creeping coprosma
Gazania rigens leucolaena	trailing gazania
Lippia canescens	lippia
Osteospermum fruitcosum	trailing African daisy
Santolina spp.	lavender cotton
Vinca spp.	periwinkle

Source: Lubin and Shelley 1997.

Sensitive and Tolerant Species

Table 5.17 gives the fuel volume at maturity for selected plants. A plant with a low fuel volume will still burn but it produces a lower-intensity fire than a plant with a high fuel volume. Trees and shrubs that contain high concentrations of volatile hydrocarbons may ignite readily and produce high-intensity fire. These include many conifer, manzanita, and eucalyptus species. Recent evidence suggests that tree maintenance, form, geometry, and moisture content may be more important than volatile hydrocarbon content.

Succulent groundcovers such as ice plant (*Lampranthus* spp., *Delosperma* spp., *Carpobrotus edulis*, and *Aptenia cordifolia*) tend to retain water during dry periods and have a greater degree of fire resistance in nonirrigated landscapes. If these groundcovers dry out, however, they can ignite. Landscape irrigation increases fire resistance.

Figure 5.87. A fire in the dumpster under this pine (*Pinus* sp.) spread to the canopy, causing the trunk and branches to be charred and needles to turn brown.

Remedies

Little can be done after thermal injury from fire has occurred. Preventive measures include:

- Select "fire-resistant" plants.
- Design landscapes to minimize fire potential. Avoid "fire ladders" (progressively taller plants that lead fire to the crown).
- Maintain well-hydrated plants through irrigation and good cultural care.
- Prune lower limbs of trees to provide a fuel break between the ground and the canopy.
- Prevent large branches from overhanging buildings, especially near chimneys.
- Remove dead branches to reduce fuel volume.
- Reduce litter buildup under trees.

Figure 5.88. Blackening or char on the trunk is a strong indicator of fire injury. Removal of a small section of the outer bark may be needed to assess the extent of damage.

Figure 5.89. If fire-injured bark becomes dry, checks, and lifts off the trunk, it is likely dead.

Scorched trees sometimes refoliate within a few months after fire damage. It is important to wait before removing apparently dead plants or plant parts. For more information on steps to take after wildfire, see *Recovering from Wildfire* (Kocher, Harris, and Nakamura 2001).

Steam and Heat

Steam released from pipes or heat released from vents, boilers, furnaces, paving equipment, or other sources can damage nearby plants (fig. 5.90). Although these types of injury are not common in landscapes, they do occur and should not be overlooked when diagnosing disorders.

Occurrence and Symptoms

Aboveground heat releases in localized areas cause scorching of leaves and dieback of stems closest to the thermal source (fig. 5.91). Releases of greater magnitude can injure the entire canopy. Usually, the onset of symptoms is rapid. Unlike fire injury, stems are not charred.

Steam or heat releases in the root zone may damage part or all the root system. Injury to a substantial portion of the root system causes extensive dehydration of leaves and stems (fig. 5.92). In cases where fewer roots are damaged, aboveground symptoms may range from canopy dieback to little or no observable injury. Such cases may prove difficult to diagnose.

It is common in container stock for soil temperature to rise above 100°F (37°C) in the summer, and root injury can occur at

Figure 5.90. Thermal injury can be caused by heat releases from equipment. Heat released from paving equipment injured trees along this street. Note that only the part of the canopy closest to the heat was injured.

Figure 5.91. Heat released from vents can cause thermal injury. The canopy of this silver dollar gum *(Eucalyptus polyanthemos)* was injured (A) by heat released from a large vent located under and next to the canopy (B). The part of the canopy closest to the vent was most severely injured.

temperatures above 105°F (40°F) (Morris and Devitt 2001). Damage is usually greatest on the sun-exposed side of the container. Root loss causes water deficit symptoms in the canopy.

Heat released from microbial decomposition of organic matter can produce root-killing temperatures. This level of heat is common in compost piles (fig. 5.93).

Look-Alike Disorders

Injury from belowground heat sources may appear similar to disorders that cause extensive root loss, such as acute water stress, mechanical damage to roots, root pathogen injury, gas injury, or chemical or herbicide injury.

Aboveground injury is usually distinctive because it is often limited to a portion of the canopy and is associated with a nearby heat source. Possible look-alike disorders include herbicide injury, foliar pathogens, and chemical injury.

Diagnosis

- Look for heat sources. Are there nearby vents, furnaces, or other heat sources? Are there underground steam pipes?
- Was equipment used in the vicinity that released significant amounts of heat, such as paving equipment?
- Does the root zone contain high quantities of organic matter, such as that found in compost piles or highly amended soils?
- For container stock, is the root distribution uniform or one-sided? Is the container directly exposed to sun?
- Did symptoms appear over a relatively short time (days or a week)?
- Are symptoms localized in confined areas or spread throughout the canopy?

Figure 5.92. Steam released from underground pipes can injure roots. This coast redwood *(Sequoia sempervirens)* was killed when a pipe containing steam broke, producing lethal temperatures in the root zone.

Figure 5.93. Although not common, plants grown in media containing high levels of organic matter may sustain root injury from heat released during the decomposition of organic matter. These California fan palm *(Washingtonia filifera)* were temporarily transplanted into a large compost pile and subsequently showed considerable dieback of fronds.

Figure 5.94. Monterey pine *(Pinus radiata)* is considered to be sensitive to high temperature injury and many decline and die within 15 years of being planted in warm summer locations. High temperatures are believed to have contributed to the decline of this young Monterey pine in Modesto, California.

Sensitive and Tolerant Species

Most species are sensitive to injury from steam or heat sources. When critical temperatures for tissue damage are exceeded, plants will be damaged regardless of species.

Certain coastal species have been observed to perform poorly when planted in hot summer locations. For example, Monterey pine *(Pinus radiata)* is noted to have a short lifespan (less than 15 years) in the Sacramento and San Joaquin Valleys (fig. 5.94). Although the heat source is the sun, this can be considered to be a form of thermal injury.

Remedies

Prevention is the best remedy for thermal injury. Avoid placing plants next to high-temperature sources. Avoid using equipment that releases excessively hot air in the vicinity of plants. If steam pipes are located in the root zone, check soil temperatures periodically to determine whether leaks have occurred.

High and Low Light

Figure 5.95. Chlorosis of this kaffir lily *(Clivia miniata)* leaf (left) resulted from high light exposure. The leaf on the right was located in the shade.

Figure 5.96. Cast iron plant *(Aspidistra elatior)* leaves become bleached when exposed to high light. This plant was exposed to afternoon sun.

Plants vary in response to light intensity: certain species are adapted to low light conditions, while others perform well in high light environments. When exposed to light levels above or below the natural range for the species, plants may be injured.

High Light Injury

Symptoms

In species that are sensitive to high light levels, chlorophyll is photooxidized in epidermal and upper palisade cells as light intensity increases above a critical level (Treshow 1970). The critical level for injury varies with species, acclimatization, and maturity of foliage. Leaves appear chlorotic or bleached (figs. 5.95–96), and in some cases necrotic areas may develop. Unlike sunburn damage, high light damage can take place when air temperatures are relatively low (see the sidebar "Sunburn, Sunscald, High Temperature Injury, and High Light Injury: What's the Difference?" p. 142).

Occurrence

Species that are intolerant of high light intensities (full sun) are readily damaged when planted without shade protection or when shade protection is removed (fig. 5.97). Poor matches of species and location are a common source of injury. In some cases, species that grow well in a location receiving morning sun may show injury when exposed to afternoon sun. The reflective surfaces of cars and buildings can also contribute to injury, as can pruning or removing a tree or tall shrub that shades light-sensitive species.

Look-Alike Disorders

High light injury is relatively distinct. In some cases, it may be confused with water deficit, aeration deficit, or nutritional deficiency. Although high light produces chlorosis symptoms similar to sunburn, leaf dehydration and necrosis are not typically associated symptoms.

Diagnosis

- Identify the species and refer to plant manuals for information on tolerance or sensitivity to high light conditions.
- Evaluate the planting location and assess the potential for high light exposure.
- Determine daily and seasonal shade patterns.
- Look for reflective surfaces.
- Determine whether any changes in shade protection have occurred.
- Examine the pattern of injury. If an injured leaf is partially shaded by another

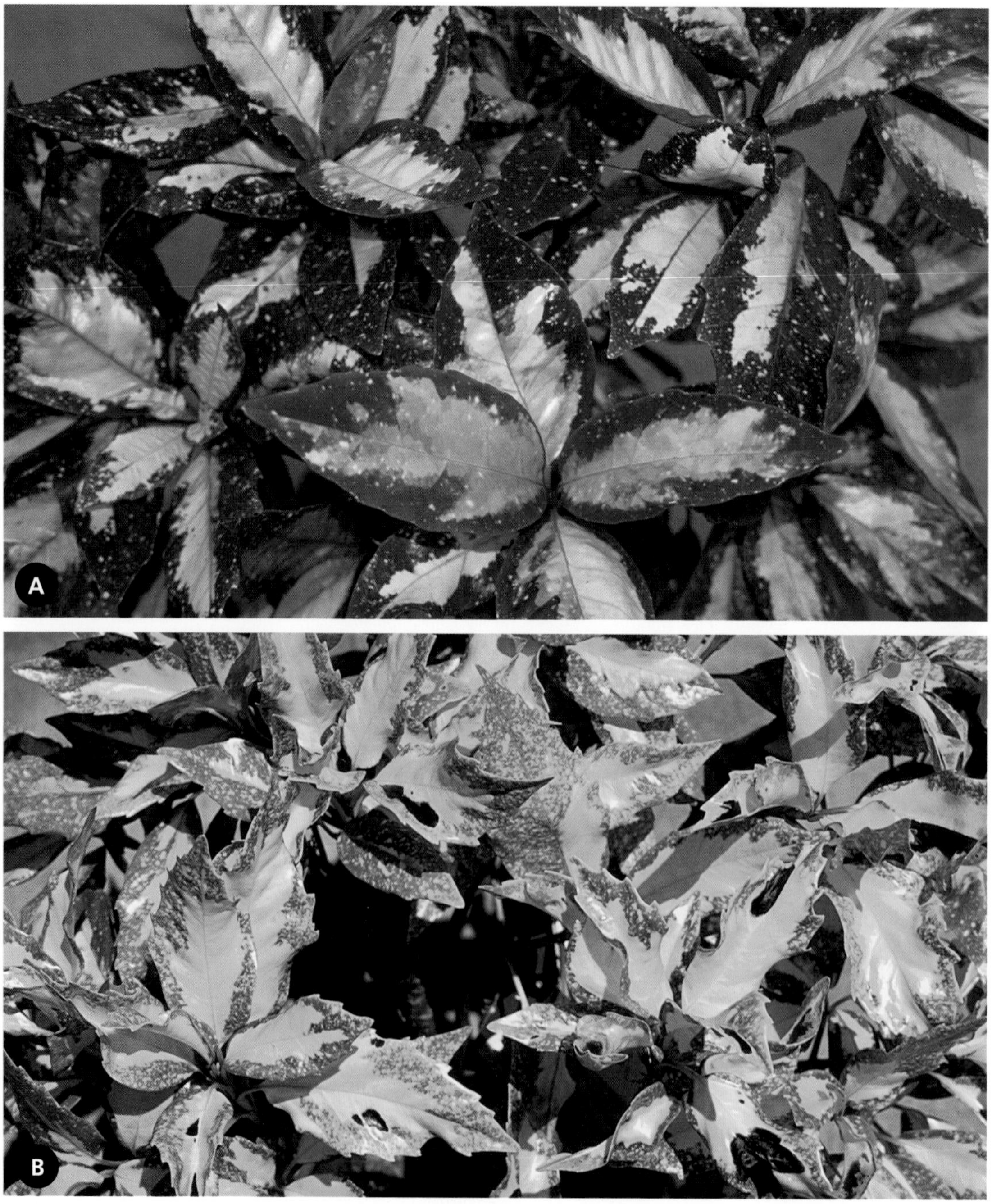

Figure 5.97. Japanese aucuba *(Aucuba japonica)* is a shade-adapted plant with dark green leaves and yellow variegation (A). When the same plant was exposed to high light conditions (full sun), leaves became bleached with necrotic lesions (B).

leaf, the shaded part will not show high light injury symptoms; leaves on the outermost part of the plant canopy will show more injury than interior leaves.

Sensitive and Tolerant Species

Table 5.18 lists landscape plants that are sensitive to high light conditions. Several plant manuals give information on the light requirements (Brenzel 2001; Brickell and Elsley 1992; Pittenger 2002). Many tropical forest and understory species are adapted to filtered light or shaded conditions and cannot tolerate full sun or high light sites. Also, species tolerant of high light conditions in coastal locations may be injured when planted in similar conditions in interior locations.

TABLE 5.18. Plants sensitive to high light conditions

Scientific name	Common name
Acer circinatum	vine maple
Aspidistra elatior	cast-iron plant
Aucuba japonica	Japanese aucuba
Caladium spp.	caladium
Clivia miniata	Kaffir lily
Fatsia japonica	Japanese aralia
Sarcococca spp.	sarcococca
Vinca minor	dwarf periwinkle

Figure 5.98. Stem elongation (etiolation) is an indicator of low light conditions in many species. In jasmine *(Jasminum polyanthum)*, internodes of stems in shade (right) are almost twice as long as stems in full sun.

Remedies

Select plants that are adapted to the planting site light conditions. Learn sun and shade patterns for the location. Plant sensitive species in sun-protected locations. Structures such as fences, screens, and shade cloth can be installed to provide shade. Overstory plants can be planted to provide shade for sensitive plants. If possible, modify or eliminate reflective surfaces that contribute to injury.

Low Light Injury

Symptoms

Low light intensities can cause subtle changes in plants. Typically, stems become elongated (etiolated), giving plants a taller, spindly, or rangy appearance (fig. 5.98). Individual leaves are larger and thinner than normal (fig. 5.99), and may exhibit a deeper green appearance than leaves in a higher light condition. Heavily shaded leaves become chlorotic and abscise (fig. 5.100). The parts of a plant canopy in the greatest amount of shade die back first. Leaves on the interior or lower portion of a canopy often abscise due to low light conditions. Flowering may be reduced or eliminated in some species.

Occurrence

Low light conditions can be found in the understory of trees, on the north side of buildings, in interior courtyards, under structures such as porch roofs or awnings, and in dense plantings. Typically, light intensity in full sun during the summer is about 12,000 foot-candles, whereas the light intensity on the shady side of a two-story building may be 2,000 foot-candles in summer and 500 foot-candles in winter.

Figure 5.99. Shaded leaves are often larger and thinner than leaves exposed to higher light levels. Here, shaded leaves of Algerian ivy *(Hedera canariensis)* (left) indicate low light conditions, as compared to leaves exposed to full sun (right).

Figure 5.100. Leaf drop and loss of canopy density on this lemon tree (*Citrus* sp.) resulted from being placed in the shade of an adjacent larger tree (A). A lemon (on left) of similar size at planting, but positioned in high light conditions, retained leaves and exhibited much greater canopy density (B).

Look-Alike Disorders

Plant growth regulators that stimulate stem elongation may produce symptoms similar to low light. Poor aeration due to excess water in the root zone may cause older or interior leaves to abscise and appear as low light injury.

Diagnosis

- Identify the species and refer to plant manuals for information on light requirements.
- Evaluate the sun and shade patterns for the location.
- Use a light meter to measure light levels (fig. 5.101).
- Compare internode length and leaf size with representative samples of the same species in a location with higher light intensity.

Sensitive and Tolerant Species

Table 5.19 lists landscape plants that tolerate shaded to mostly shaded sites. In general, different types of landscape plants require different light intensities (see also fig. 5.102):

• most woody plants (full sun: winter) and summer)	5,000–12,000 foot-candles
• shade plants	1,000 foot-candles
• house plants	50–500 foot-candles

Several plant manuals give information on the light requirements of landscape species; see, for example, Brenzel 2001, Brickell and Elsley 1992, and Pittenger 2002.

Remedies

Select plants that are adapted to light conditions at the site. If appropriate, reduce the amount of shade by pruning overstory trees or shrubs or by removing shading structures. In some cases, it may be possible to transplant or move the plant into a higher light condition.

TABLE 5.19. Plants tolerant of shaded to mostly shaded sites

Scientific name	Common name
Acer buergeranum	trident maple
Acer campestre	hedge maple
Acer palmatum	Japanese maple
Caryota mitis	fishtail palm
Chamaecyparis nootkatensis 'Pendula'	Nootka cypress
Chionanthus retusus	Chinese fringe tree
Cornus florida	flowering dogwood
Howea forsterana	paradise palm
Ilex opaca	American holly
Ilex vomitoria	yaupon holly
Podocarpus macrophyllus	yew pine
Podocarpus nagi	broadleaf podocarpus
Prunus caroliniana	Carolina laurel cherry
Taxodium distichum	bald cypress
Taxus baccata	English yew
Thuja plicata	western red cedar
Trachycarpus fortunei	windmill palm
Viburnum odoratissimum	sweet viburnum

Source: Gilman 1997.

Estimating Light Intensity Using a Camera with Built-In Light Meter

To estimate the intensity of sunlight falling on a plant using a camera with a built-in light meter and whose film and shutter speed can be manually adjusted:

- Set film speed to ASA 25.
- Set shutter speed to 1/60.
- Hold a sheet of opaque white paper next to plant.
- Point camera at paper at a distance equal to the narrow width of the paper.
- Adjust lens opening (f-stop) to give the proper exposure as indicated by the built-in meter.

f-stop	Light intensity (foot-candles, approx.)
2	40
2.8	75
4	150
5.6	300
8	600
11	1,200
16	2,400

Note: Photometers are not accurate for measuring light from most artificial sources.

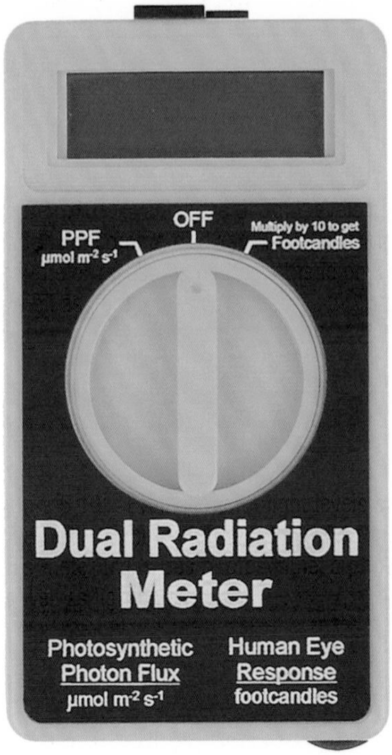

Figure 5.101. Evaluations of light levels can be made with commercially available light meters. This meter measures photosynthetic photon flux as well as light intensity.

Figure 5.102. Under low light conditions (<100 foot-candles), the waxleaf privet *(Ligustrum japonicum)* on right has etiolated stems and is smaller than the one grown under higher light (5,000 foot-candles) conditions (left). The plant in the middle was grown in moderate shade (300 to 600 foot-candles).

Wind

Wind can desiccate leaf and bark tissues and physically break branches, foliage, and flowers. A gradual change in tree form is common in coastal or other particularly windy areas (fig. 5.103).

Figure 5.103. Coastal winds can cause growth of Monterey cypress *(Cupressus macrocarpa)* to be asymmetrical (A) and that of shore pine *(Pinus contorta)* and cypress (*Cupressus* sp.) to be dwarfed or compressed (B).

Symptoms

Wind commonly causes water deficits and associated injury symptoms. Wind can increase plant water loss by as much as 30 percent. If soil moisture levels are low, or if water absorption and transport do not match transpirational water loss, water stress follows. If water loss is severe, leaves develop marginal necrosis and may abscise prematurely (figs. 5.104–105). Leaf necrosis is particularly severe during periods of hot, dry winds. In less-severe conditions, leaves may be smaller than typical for the species. As a result of water deficits and early abscission, the canopy of wind-damaged plants becomes one-sided, with most foliage on the leeward side (fig. 5.106).

Strong winds can break branches, leaves, and fruit. Typically, leaves remain attached to the tree, but they may tear or shred (fig. 5.107).

In desert areas where fine sand is picked up by wind and blown against plants, abrasion of tissues can occur. The impact of sand grains on leaf, branch, and trunk surfaces causes them to appear grazed or sandblasted. In addition, wind-blown sand can bury the root crown of trees and shrubs.

Occurrence

Wind speed, duration, and direction, and plant exposure determine the extent of wind damage. Generally, wind velocities above 30 mph (48 km/hr) are needed to break plants, but injury can be found at lesser wind speeds. Plants located in particularly windy areas are prone to injury: coastal areas, tops of hills or mountains, and wind tunnels (urban areas where wind is channeled by buildings or other structures). The hot, dry Santa Ana winds in Southern California and the cold, dry north winds of the Central Valley commonly cause plant injury. In areas where wind patterns change (e.g., where a building or a row of trees has been removed), previously protected plants can be damaged by wind. Container plants, especially trees, placed in windy locations are particularly sensitive to injury.

Look-Alike Disorders

Wind injury and water deficits are closely related, and the foliar symptoms are similar. In addition, salt injury, specific ion toxicity, herbicide injury, and insect injury (thrips or mites) can resemble wind injury.

Diagnosis

- Identify the plant and refer to plant manuals for information regarding the wind tolerance or sensitivity of the species.

Figure 5.104. The canopy of this deciduous oak (*Quercus* sp.) has become very thin and asymmetrical because of wind (A). Leaves have become discolored and tattered, and exhibit marginal necrosis (B).

Figure 5.105. Wind can cause leaves to abscise prematurely. These purple-leaf plum (*Prunus cerasifera* 'Atropurpurea') are planted in a windy location and defoliate by midsummer (A). Nearby purple-leaf plums are protected by houses and have full canopies (B).

Figure 5.106. Winds channeling over rooftops has caused the canopy of this London plane *(Platanus × acerifolia)* to become deformed. Growth occurred preferentially on the leeward side.

- Look for damage patterns: wind damage will be most severe on the side of plants facing the prevailing wind.
- Learn the history of wind events in the location. Have there been strong winds in the area? Wind data can be obtained from the California Irrigation Management Information System (CIMIS), the UC IPM Web site (www.ipm.ucdavis.edu), or local weather stations.
- Are symptoms consistent with those described for wind injury?
- Is the plant in a particularly windy area?
- Is the soil dry? Is the plant irrigated?

Sensitive and Tolerant Species

Wind-tolerant species often have foliage that is narrow or needlelike and covered with a thick cuticle (table 5.20). Broadleaf plants with thin leaves (e.g., Japanese maple, *Acer palmatum*) or large leaves (e.g., London plane, *Platanus acerifolia*) are subject to wind damage. Refer to plant manuals for information regarding the wind tolerance of species.

Remedies

Select wind-tolerant species for windy locations. Plant sensitive species in wind-protected sites only. Some species can be used as windbreaks for structures or other plants (table 5.21), though they may not be

Figure 5.107. Wind can cause leaves to be shredded, torn, or tattered. These banana (*Musa* sp.) leaves are shredded (A), while London plane *(Platanus × acerifolia)* leaves are tattered and torn (B).

TABLE 5.20. Plants tolerant of windy environments

Scientific name	Common name
Acacia longifolia	Sydney golden wattle
Arctostaphylos insularis	Island manzanita
Arctostaphylos manzanita	common manzanita
Atriplex lentiformis	quailbush
Berberis darwinii	Darwin barberry
Buxus sempervirens	English boxwood
Callistemon citrinus	lemon bottlebrush
Carissa macrocarpa	Natal plum
Cassia artemisioides	feathery cassia
Ceanothus arboreus	feltleaf ceanothus
Ceanothus thyrsiflorus	blue blossom
Chamaecyparis lawsoniana	Lawson's cypress
Cistus spp.	rockrose
Coprosma × *kirkii*	creeping coprosma
Coprosma repens	mirror plant
Cotoneaster (various species, not all)	cotoneaster
Dodonaea viscosa 'Purpurea'	clammy hopseed
Eleagnus pungens	silverberry
Eriogonum arborescens	Santa Cruz Island buckwheat
Eriogonum giganteum	St. Catherine's lace
Escallonia spp.	escallionia
Euonymus japonica	evergreen euonymus
Garrya elliptica	coast silktassel
Griselinia littoralis	broadleaf kapuka
Griselinia lucida	puka
Hakea laurina	pincushion tree
Ilex aquifolium	English holly
Leptospermum laevigatum	Australian tea tree
Leptospermum scoparium	New Zealand tea tree
Ligustrum spp.	privet
Melaleuca spp.	melaleuca
Myrica californica	Pacific wax myrtle
Myrsine africana	African boxwood
Myrtus communis	true myrtle
Phormium tenax	New Zealand flax
Pittosporum spp.	pittosporum
Prunus ilicifolia	hollyleaf cherry
Prunus laurocerasus	English laurel
Pyracantha spp.	pyracantha
Rhamnus alaternus	Italian buckthorn
Rhamnus californica	coffeeberry
Rhaphiolepis indica	India hawthorn
Rhus integrifolia	lemonade berry
Rhus ovata	sugar bush
Rosmarinus officinalis	rosemary
Tamarix spp. (avoid *T. chinensis*)	tamarisk
Xylosma congestum	xylosma

TABLE 5.21. Trees useful as landscape windbreaks

Scientific name	Common name
Calocedrus decurrens	incense cedar
Cupressus arizonica	Arizona cypress
Dodonaea viscosa	hopseed bush
Elaeagnus angustifolia	Russian olive
Elaeagnus pungens	silverberry
Escallonia spp.	escallonia
Hakea spp.	hakea
Leptospermum laevigatum	Australian tea tree
Ligustrum lucidum	glossy privet
Myoporum laetum	myoporum
Picea abies	Norway spruce
Pinus canariensis	Canary Island pine
Rhamnus alaternus	Italian buckthorn
Taxus baccata 'Stricta'	Irish yew
Thuja occidentalis	American arborvitae
Thuja plicata	western red cedar

suitable for small residential areas; wind screens can be used to protect young trees during the establishment period (fig. 5.108). Protecting larger plants can be difficult because protective structures must be large and durable. When removing a group or row of trees or large shrubs, consider the potential for wind exposure for remaining vegetation.

Figure 5.108. Sensitive plants can be protected from wind. Here, young plants are being protected by a row of mature blue gum *(Eucalyptus globulus)* (A), and a wind screen is used to protect young Monterey cypress *(Cupressus macrocarpa)* (B).

Gas Injury

Gases released into the plant canopy or root zone can cause injury. The gases may be directly phytotoxic to plant parts, or they may exclude oxygen from the root zone. Leaks from gas lines, sewer lines, vents, and landfills are relatively common sources of gas injury.

Figure 5.109. Gases released into the root zone of plants can cause significant injury. Dieback of this Canary Island pine *(Pinus canariensis)* (A) resulted from a leak in a nearby gas line (B).

Symptoms

Gas injury may be lethal or sublethal depending on the type of gas, amount released, soil permeability, duration of exposure, sensitivity of the species, and condition of the plant. Excessive or prolonged releases are usually lethal. Slow releases may result in symptom development over several months or years.

Gas Release in Soil

Phytotoxic gas release in the root zone causes loss of root function and, frequently, root death. Affected plants have a diminished capacity to absorb water and nutrients, and plant growth and development are reduced. Symptoms range from slow growth to death of the whole plant (fig. 5.109). Initially, plants show symptoms similar to water stress: slow growth, small leaves, wilt, and leaf drop. Only partial leaf-out may occur in the spring. Roots may appear water-soaked and bluish, while the soil may turn bluish gray or black. As the injury becomes more severe, a few branches or the entire canopy may die, and the whole plant may eventually be killed.

Atmospheric Gas Release

The release of phytotoxic gases in or around a plant's canopy usually causes severe injury, and the onset of symptoms is rapid. The parts of the canopy closest to the gas source show the highest levels of injury. If gas release is sufficient in amount and duration, the entire canopy is affected (fig. 5.110). Ammonia gas release causes a bleaching, bronzing, or blackening of the foliage, depending on plant species. Chlorine gas causes marginal chlorosis, stippling of the leaf upper surface, reddening, and necrosis. Pine needles develop a bright reddish brown necrosis from the tip inward, and a small band of stippling may be found below the necrotic area. Industrial gases that are strong oxidizing or reducing agents are likely to be phytotoxic.

Occurrence

Gas Release in Soil

In the root zone, gas injury typically is caused by a gas-line or sewer-line break that releases methane gas, or by gases leaking from landfills; these gases consist primarily of methane and carbon dioxide, with traces of carbon monoxide, ammonia, ethane, hydrogen sulfide, ethylene, propylene, and hydrogen cyanide.

Over the last 30 years, many parks, golf courses, and playgrounds have been de-

Figure 5.110. Plant injury may result from atmospheric release of phytotoxic gases. Dieback in this London plane *(Platanus × acerifolia)* resulted from gases released from nearby industrial equipment.

Figure 5.111. Gases released in landfill sites can cause injury to many species. Typically, injury is acute and severe. So far, trees planted on this landfill have not been injured.

veloped on landfills. Although some plant species have higher levels of tolerance than others, most species are injured by landfill gases (fig. 5.111). Depending on the source, gas levels may build up to injurious levels either rapidly or slowly. Gas line leaks are usually rapid, while landfill releases are often gradual.

Although some soil-released gases may be phytotoxic, others injure plants by excluding oxygen from the root zone. For example, methane and carbon dioxide released in landfills act to displace oxygen, creating anaerobic conditions.

Atmospheric Gas Release

Occasionally, gases released from vents, chimneys, smokestacks, or industrial equipment (e.g., refrigeration units) injure plants. Although many atmospherically released gases are innocuous (e.g., nitrogen), some are phytotoxic (e.g., ammonia). Government regulations have limited emissions of industrial gases that may also be phytotoxic. Phytotoxic gas releases are usually accidental and are typically short-lived.

Look-Alike Disorders

Gas injury may be mistaken for water deficit, aeration deficit, root disease, herbicide injury, or air pollution.

Diagnosis

Gas Release in Soil

- Smell the soil in the area around the injured plant. Do you detect gas odors that are not typical? Keep in mind that some gases cannot be detected by smell either because of their chemical composition or because their concentration is low.
- Inspect the soil. Has the color turned blue, bluish gray, or black? These colors indicate the presence of ions in a reduced state and, therefore, anaerobic conditions.
- Inspect plants in the area. Are all species affected? Or only one species? For most gas leaks, more than one species will be affected.
- Determine whether gas lines are in the vicinity. This may require assistance from a utility line detection service.
- Determine whether the landscape has been developed on landfill. If so, determine the depth of landscape soil, type and thickness of the cap layer, and the type of landfill materials. Has the soil settled in some areas? Are there cracks in the soil?
- Use gas-detecting equipment to determine levels of potentially injurious gases in the root zone. Combustible gases, hydrogen sulfide, carbon dioxide, and many others can be detected with specialized instruments. If instruments are not available, collect samples for laboratory analysis.

Atmospheric Gas Release

- Examine the area around the injured plant(s). Are there obvious potential sources of potentially phytotoxic gas (e.g., vents, chimneys, smokestacks, industrial equipment)? What gases may have been released? When were releases made? When did injury occur?
- Is only one species affected? In most phytotoxic gas releases, most species in the exposed area will be injured.
- Is injury greatest in plants or plant parts that are nearest to the release location?

Sensitive and Tolerant Species

Generally, shallow-rooted species are more tolerant of gases produced in landfill conditions than are deep-rooted species. Table 5.22 is a list of species tolerant to landfill conditions, ranked in order of tolerance; table 5.23 lists species sensitive to chlorine gas.

Remedies

- Once the location of a gas leak has been identified, shut off the source and allow gases to vent from the area. Monitor gas levels until they return to acceptable levels. Fix the leak when ambient concentrations pose no threat to workers.

- In landfills, locate and stop the leak (if possible). If this cannot be done, plant species considered tolerant of landfill sites.
- Maintain adequate moisture levels in the root zone.
- For atmospheric releases, attempt to prevent releases in the landscaped area or redirect gases to areas where plants will not be injured (e.g., above the plant canopy).
- For soil around natural gas-line leaks, leave the excavation pit open for 1 to 4 weeks to allow for dissipation of the gas. Then refill the pit.

TABLE 5.22. Plants tolerant of landfill conditions

Scientific name	Common name
Acer rubrum	red maple
Euonymus alata	winged euonymus
Fraxinus pennsylvanica var. *lanceolata*	green ash
Ginkgo biloba	ginkgo
Gleditsia triacanthos	honey locust
Liquidambar styraciflua	sweetgum
Myrica pensylvanica	bayberry
Nyssa sylvatica	tupelo
Picea abies	Norway spruce
Pinus strobus	white pine
Pinus thunbergii	Japanese black pine
Platanus occidentalis	American sycamore
Populus spp.	mixed poplar
Populus spp.	hybrid poplar
Quercus palustris	pin oak
Rhododendron roseum elegans	rhododendron
Salix babylonica	weeping willow
Taxus cuspidata 'Capitata'	Japanese yew
Tilia americana	American basswood

Source: Leone et al. 1980.

TABLE 5.23. Plants sensitive to chlorine gas

Scientific name	Common name
Aesculus spp.	buckeye
Catalpa bignonioides	Southern catalpa
Chrysanthemum morifolium	chrysanthemum
Medicago sativa	alfalfa
Petunia hybrida	petunia
Picea pungens	blue spruce
Pinus contorta murrayana	lodgepole pine
Pinus sylvestris	Scotch pine
Raphanus sativus	radish
Taraxacum officinale	dandelion

Source: Treshow 1970.

Air Pollution

Air pollutants recognized as being phytotoxic include ozone (O_3), sulfur dioxide (SO_2), and peroxyacetyl nitrate (PAN). In California, injury from ozone is most common. Prior to 1960, PAN caused injury in the Los Angeles basin, but now levels are low and injury is rare. Also, sulfur dioxide levels have declined and injury is less common.

Ozone

Ozone is produced by the reaction of volatile hydrocarbons and nitrogen oxides (mainly from vehicle emissions) in the presence of sunlight. Elevated ozone concentrations are a problem in summer, but not in winter, when lower temperatures slow atmospheric reaction rates. Entering through stomata, ozone causes cell membrane deterioration, loss of chlorophyll, and may cause degradation of cuticular wax.

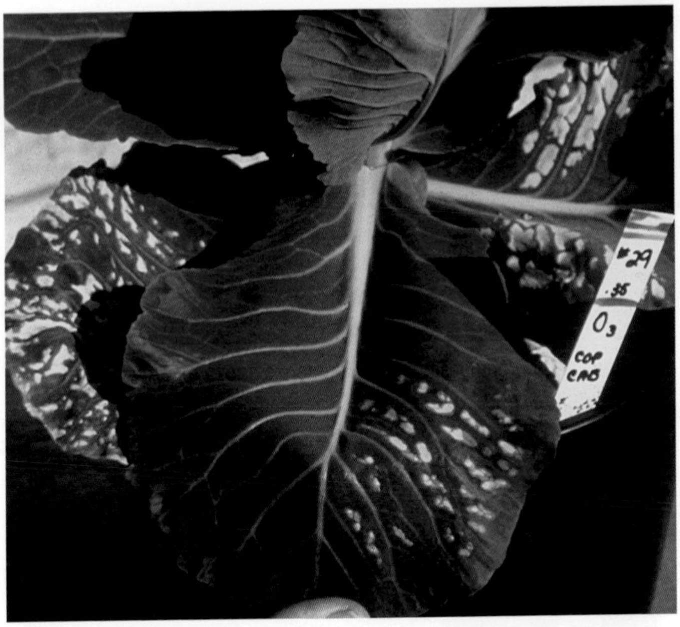

Figure 5.112. Acute ozone exposure caused bleaching on this cabbage *(Brassica oleracea* var. *capitata)* leaf.

Symptoms

Ozone can cause foliar chlorosis, bleaching, flecking, mottling, stippling, and necrosis. Damage varies with species, cultivar, weather conditions, and age of plant. Generally, symptoms appear on the upper surface of leaves. Variable symptoms can occur on the same plant or on an individual leaf.

Injury can be chronic or acute. Chronic injury results from long-term exposure to relatively low concentrations, while acute injury results from short-term exposure to higher concentrations. Typical summertime maximum ozone concentrations (ambient surface measurements) in polluted urban or suburban areas range from about 0.1 ppm (parts per million) to 0.4 ppm in the most highly polluted cities of the world; in relatively unpolluted rural areas levels are about 0.03 ppm. Sensitive plants have been observed to exhibit visual symptoms at levels of 0.08 ppm or lower.

Acute Injury

Broadleaf plants

- Bleaching: Upper leaf surface bleaching appears either generally or as small necrotic spots (fig. 5.112).
- Flecking: Small spots that are metallic or brown fading to tan, gray, or white on leaves. (fig. 5.113).
- Stippling: Smallish white, black, red, or reddish purple spots on leaves.
- Necrosis: Dead tissue extending all the way through the leaf or flower; leaf turns white to dark orange-red and larger veins may remain green. Flower tissues turn brown or black (fig. 5.114).

Conifers

- Needle banding: Yellow (chlorotic) bands develop on semimature needle tissue.

- Tip burn: Brown needle tips on young elongating needles. May first appear as reddish brown areas that progress to brown.

Chronic Injury

Broadleaf plants

- Bronzing: Leaves turn reddish brown to brown.
- Chlorosis: Yellowing of foliage (figs. 5.115–116).
- Premature leaf drop (senescence): Early loss of leaves, flowers, or fruit.

Conifers

- Flecking: Earliest symptom on older needles.
- Mottling: General diffuse yellow areas mixed with green sections on first-year needles.

Figure 5.113. Flecking or stippling on cotton *(Gossypium hirsutum)* was caused by acute ozone exposure. Uninjured leaves on left.

Figure 5.114. Ozone exposure caused necrosis on this orchid (*Cattleya* sp.) flower. Uninjured flower on left.

- Premature senescence: Early needle drop, usually within 2 years after developing.

Occurrence

Areas of California with elevated summer ozone levels include portions of the Los Angeles Basin, the San Francisco Bay Area, the San Diego area, and parts of the Central Valley, especially around Sacramento, Fresno, and Bakersfield. In addition, injury has been found on native vegetation in the Sierra Nevada and San Bernardino mountain ranges adjacent to the air basins mentioned above and below the atmospheric boundary layer (about 6,000 feet, or 1,830 m).

Generally, young plants are more sensitive than older ones. Mature leaves with fully functional stomata are most sensitive. Temperature, light level, irrigation, and fertilization can affect the level of injury.

Look-Alike Disorders

- Foliar damage caused by mites (stippling) or thrips (bronzing).
- Foliar diseases such as anthracnose can produce necrotic spots, interveinal necrosis, or chlorosis in many species.
- Herbicide injury.

Diagnosis

Collect information regarding air quality of the site from air quality management districts, local weather records, or newspaper weather pages. Determine whether potentially phytotoxic levels of air pollutants occurred when injury symptoms were noticed. Ozone injury symptoms may take a few days to develop after acute exposure. Many plants growing in high-ozone areas show symptoms of chronic exposure that are mistaken as normal for lack of comparison with plants gown under low-ozone levels.

- Identify the plant and determine whether it is considered to be a sensitive species.
- Are other species affected?
- Was onset of symptoms rapid or slow?
- Are symptoms on the upper surface of the leaf, lower, or both?

Figure 5.115. Chlorosis on this mother fern *(Asplenium bulbiferum)* was caused by chronic ozone exposure.

Figure 5.116. Yellow mottling and browning of Japanese maple *(Acer palmatum)* leaves was caused by ozone exposure.

Figure 5.117. Whitish interveinal necrotic areas on these petunia *(Petunia hybrida)* leaves were caused by sulfur dioxide exposure.

Sensitive and Tolerant Species

Plants with thin leaves and cuticles are more likely to be injured by ozone than plants with sclerophyllous (hard and stiff) leaves (table 5.24). Sensitive species may have genetic variants (ecotypes, varieties, cultivars) that have greater tolerance to ozone. For example, the 'Santa Ana' hybrid of bermudagrass developed by the University of California is more smog-tolerant than hybrid bermudagrasses that were developed in areas of Georgia with little ozone pressure.

Remedies

Select tolerant species for areas prone to phytotoxic levels of ozone.

Sulfur Dioxide

Power and manufacturing plants that burn high-sulfur coal or oil typically emit sulfur dioxide. Recent advances in emissions control technology, however, have substantially reduced or eliminated sulfur dioxide emissions. As a result, plant injury caused by sulfur dioxide is no longer common in the United States, especially on the west coast (high-sulfur coal is a product of mines in the eastern United States).

Symptoms

Although sulfur is an essential plant nutrient, sulfur dioxide reduces chlorophyll content and may damage chloroplasts, reducing photosynthesis and growth. Symptoms vary among broadleaf (dicot), narrowleaf (monocot), and conifer species.

Dicots

Acute symptoms begin as dull green or water-soaked areas between major leaf veins, which turn whitish or light gray (fig. 5.117). Interveinal bleaching or yellowing occurs under chronic exposure conditions . Some leaves may have a silvery appearance.

Monocots

Yellowish white or ivory streaks develop near leaf tips and extend down between the veins. Leaf margins commonly turn brown.

Conifers

Initially, reddish brown necrotic needle tips are found (fig. 5.118) that may develop into distinct bands if successive acute episodes occur (fig. 5.119). Premature needle drop is common. Older needles begin to yellow from the tip downward in cases of chronic exposure; needle drop may follow. Young needles appear to be normal.

TABLE 5.24. Plants sensitive to ozone

Scientific name	Common name
Abutilon spp.	flowering maple
Acer negundo	box elder
Acer saccarinum	silver maple
Asclepias syriaca	milkweed
Alnus spp.	alder
Bauhinia spp.	orchid tree
Calycanthus occidentalis	western spicebush
Ceratonia siliqua	carob tree
Chrysanthemum spp.	chrysanthemum
Cinnamomum camphora	camphor tree
Crataegus spp.	hawthorn
Fraxinus spp.	ash
Fuchsia spp.	fuchsia
Gleditsia spp.	locust
Hibiscus spp.	hibiscus
Juglans regia	English walnut
Liquidambar styraciflua	sweetgum
Liriodendron tulipifera	tulip tree
Morus alba	white mulberry
Persea spp.	avocado
Petunia hybrida	petunia
Pinus nigra	Austrian black pine
Pinus ponderosa	ponderosa pine
Platanus acerifolia	London plane tree
Populus tremuloides	quaking aspen
Populus spp.	poplar
Ranunculus spp.	buttercup
Rhododendron spp.	azalea
Robinia pseudoacacia	black locust
Rudbeckia laciniata	cut-leaf coneflower
Salix spp.	willow
Schinus molle	California pepper
Trifolium spp.	clover (red, subterranean, white)
Ulmus spp.	elm
Vicia spp.	vetch
Vinca spp.	periwinkle

Source: Noble 1988.

Occurrence

Close proximity to sulfur-emitting industrial or refinery sites is usually essential for sulfur dioxide injury. However, local geographic features and weather conditions can move toxic concentrations to distant sites. There are few locations in California where sulfur dioxide levels have caused plant injury.

Look-Alike Disorders

Low temperature, water deficit, aeration deficit, nutrient deficiency, herbicide injury, and certain insect and disease pests can produce injury symptoms similar to sulfur dioxide.

Diagnosis

Although tissue analysis can confirm sulfur dioxide injury, laboratory reports can be difficult to interpret (see Flagler 1998). Symptoms may not be confined to one species or in one area, and multiple species in an area may show symptoms. Indicator plants such as lichens, mosses, alfalfa, hybrid poplar, and bracken fern have been used to assess sulfur dioxide injury.

Figure 5.118. Sulfur dioxide injury to pine (*Pinus* sp.) needles.

TABLE 5.25. Plants sensitive to sulfur dioxide

Scientific name	Common name
Acer saccharinum	silver maple
Betula papyrifera	white paper-bark birch
Betula pendula	European white birch
Corylus var. *cornuta*	California hazel
Parthenocissus quinquefolia	Virginia creeper
Pinus jeffreyi	Jeffrey pine
Pinus ponderosa	ponderosa pine
Pinus sylvestris	Scotch pine
Platanus occidentalis	American sycamore
Populus nigra	Lombardy poplar
Populus tremuloides	quaking aspen
Pseudotsuga menziesii	Douglas fir
Quercus macrocarpa	bur oak
Robinia pseudoacacia	black locust
Syringa vulgaris	common lilac
Thuja occidentalis	American arborvitae
Ulmus americana	American elm
Ulmus parvifolia	Chinese evergreen elm

Figure 5.119. Acute sulfur dioxide exposure caused banding on pine (*Pinus* sp.) needles.

Sensitive and Tolerant Species

Table 5.25 lists selected plants sensitive to sulfur dioxide injury.

Remedies

Typically, no remedies are needed because this is a relatively uncommon air pollutant in the western states. Avoid the use of sensitive plants in areas where sulfur dioxide may be a problem.

Using Indicator Plants To Detect Smog

Although scientists use analytical instruments to measure air pollutants very accurately, smog-sensitive plants can be used by landscape professionals as bioindicators of air pollution (Flagler 1998). "Sentinel" plants are nonnative species that show well-defined symptoms when exposed to polluted air for a short period. They are grown in clean air and then exposed to the ambient polluted air. Ozone indicator plants include annual bluegrass *(Poa annua)*, bean *(Phaseolus vulgaris)*, white clover *(Trifolium repens)*, morningglory (*Convolvulus* spp.), spinach *(Spinacia oleracea)*, and tobacco (*Nicotiana* spp.). "Detector" plants are native species that show symptoms from long-term exposure to ozone. Examples include blackberry (*Rubus* spp.), green ash *(Fraxinus pennsylvanica* var. *lanceolata)*, milkweed (*Asclepias* spp.), sassafras *(Sassafras albidum)*, and white ash *(Fraxinus americana)*.

Lightning Injury

Lightning strikes may kill trees outright or severely damage branches or portions of the trunk. These wounds and dead branches attract insects and provide infection sites for decay-causing fungi.

Symptoms

Lightning injury has immediate and long-term symptoms. An entire tree may burst into flames when struck by a "hot bolt" with temperatures of over 25,000°F (14,000°C). The trunk and branches of a tree may be completely blown apart by "cold lightning," which strikes at 20,000 miles per second (32,000 km/s). The upper trunk and branches of conifers may be killed outright, while lower portions of the tree remain unaffected; the trunk may split (fig. 5.120). In some cases, a continuous strip of bark and sapwood is blown off a branch and the trunk, leaving a groove that follows the grain of the wood from the point of the strike to the ground (fig. 5.121). The bark often hangs in shreds at the edges of the wound. Internal trunk tissues may be severely burned without external evidence. The bark and soil may be blown off a root before the charge dissipates in the soil. All or part of a tree's root system may be killed as a result of xylem injury, with no external symptoms present. In crowded groves, trees near the one directly struck may also be killed.

Figure 5.120. Lightning caused a large split in the trunk of this tree.

In the long term, a sheath of abnormal wood (a ring in cross-section) may develop beneath the bark, and in some cases, galls may develop along the trunk. If roots are killed, adventitious roots may grow from the trunk. Invasions by secondary insects or diseases often cause lightning-injured trees to decline slowly (fig. 5.122).

The severity of injury is related to the lightning intensity, the amount of water on and in the bark, the characteristics of the branch and trunk tissues, and the tree species.

Occurrence

Lightning injury is uncommon in urban areas in California. Trees most susceptible to lightning injury include lone trees in open areas, tall trees on the windward edge of groups or stands, the dominant tree in a stand, a tree growing in moist soil or near a body of water, and a tree close to a building. While all tree species may be struck by lightning, some species seem to be more susceptible (table 5.26). According to some authorities, the most susceptible trees are

Figure 5.121. A longitudinal section of bark was stripped from this tree by lightning (A). In some cases, injury extends into the wood (B).

high in starch content and are therefore good conductors of electricity; the least susceptible trees are high in oil content and are poor conductors of electricity. Also, deep-rooted and decaying trees appear to be struck more often than shallow-rooted and healthy trees.

Look-Alike Disorders

The dieback and slow decline of lightning-injured trees may be confused with other disorders that cause slow death, such as root disease, mechanical root damage, girdling injuries, and adverse soil conditions. However, immediate symptoms are distinctive and are usually not confused with other disorders.

Figure 5.122. Wood decay and insect infestation can follow lightning injury.

Diagnosis

- Consider tree height and location.
- Check weather records for recent lightning storms.
- In old, declining trees, look for the characteristic groove that follows the grain of the wood from the point of the strike to the ground.

TABLE 5.26. Trees most often and least often struck by lightning

Scientific name	Common name
Most often struck	
Acer spp.	maple
Fraxinus spp.	ash
Liriodendron tulipifera	tulip tree
Picea spp.	spruce
Pinus spp.	pine
Platanus spp.	sycamore
Populus spp.	poplar
Quercus spp.	oak
Tsuga spp.	hemlock
Ulmus spp.	elm
Least often struck	
Aesculus hippocastanum	horsechestnut
Betula spp.	birch
Fagus spp.	beech

Source: Cripe 1978.

Sensitive and Tolerant species

See table 5.26.

Remedies

Repairs to struck trees should be limited to safety pruning and cleanup. Further efforts depend on whether the tree survives and what condition it is in. Immediately after the strike, loose bark should be tacked back into place and kept moist to prevent drying. Improve vigor of lightning-injured plants with good culture, especially proper irrigation. Prune or remove dead branches to reduce potential hazards. Lightning rods with copper cables leading down the trunk and through the soil to grounding rods can be installed in trees for protection.
The National Arborist Association (http://www.nat.arb.com/) and the National Fire Protection Association (http://www. nfpa.org/) have published standards for such systems.

Hail Injury

Figure 5.123. Leaves of this styrax *(Styrax obassia)* were torn and tattered during a hailstorm in Virginia.

Figure 5.124. On fleshy leaves, hail injury may appear as grazing or pitting of the upper surface, as seen on this century plant *(Agave americana)*.

Hail principally causes leaf damage, but in severe cases, hailstones may break branches or injure the bark of trees and shrubs.

Symptoms

Symptoms of hail injury appear immediately or within a few days, depending on the size and quantity of the hailstones and the plant species affected. Symptoms usually include torn or tattered leaves (fig. 5.123) and grazing or pitting of leaves (fig. 5.124). Twigs may be broken, and bark may be bruised, broken, or scarred (fig. 5.125). Hail wounds on bark are usually elliptical and vary in length from a fraction of an inch to 4 inches (10 cm) or more. Hail wounds occur on the upper side of branches and on the side of the plant facing the storm. Severe hail wounds may coalesce and kill all the bark on one side of a stem. Twigs and branches may die if tissues dry as a result of the wounds. Bark injuries may serve as entry points for decay-causing fungi, especially if the injuries do not close rapidly. Twigs and branches are more likely to die if the injury occurs during dormancy than during the growing season. Also, hailstones that collect on or around a plant may cause low temperature injury (fig. 5.126). All species in the locality of the hail storm display similar symptoms, although the severity of symptoms varies among species.

Occurrence

Hail showers do occur in California, but damage to landscape plants is typically not found.

Figure 5.125. The tiny scars on this citrus (*Citrus* sp.) branch were caused by hail.

Look-Alike Disorders

Certain canker diseases may produce bark injuries similar to hail damage. However, such cankers occur randomly on branches, while hail damage would occur on the upper side of the branches only.

Diagnosis

Check weather records to verify that a hailstorm has occurred. Examine all plants in the landscape for symptoms.

Remedies

Improve vigor of hail-injured plants with good culture, especially proper irrigation.

Figure 5.126. Hailstones collecting at the base of the leaves of this century plant *(Agave americana)* (A) or in low-growing groundcovers (B) may cause low temperature injury to exposed tissues.

Girdling and Kinked Roots

Girdling roots encircle the trunk or other roots (fig. 5.127); kinked roots have a sharp bend of 90° or more (fig. 5.128). Both defects may affect the health and structural stability of container-grown or field-grown trees. The severity of these defects is determined by their location (in the root system) and the amount of root system that remains unaffected. Severe defects may cause tree decline or structural failure (fig. 5.129).

Occurrence

Restrictions in rooting space imposed by seed flats or containers during the plant production process can lead to kinked or girdling roots. Blocked by container sides, roots are redirected around and across the container, forming kinked or girdling roots (fig. 5.130).

Girdling roots form when lateral roots develop perpendicularly to primary roots. When arising close to the trunk or another root, the enlarging girdling root constricts the stem or root, limiting diameter and reducing xylem and phloem transport (fig. 5.131). In some cases, the injury may not be apparent until many years after planting.

Kinked roots develop in the production phase when young trees are transplanted from seed flats or nursery containers. If roots are not carefully spread out or if transplanting is delayed and roots contact the bottom of the seed flat, kinking will occur.

Figure 5.127. Girdling or circling roots may encircle the trunk or structural roots. Here, trunk development will be limited by girdling roots.

Figure 5.128. Roots with a sharp bend (90° or more) are considered to be kinked. Kinked roots may not provide sufficient structural support for larger trees. Several kinks are seen in the root system of this elm (*Ulmus* sp.)

Figure 5.129. A girdling root on this Italian stone pine *(Pinus pinea)* led to its failure during a windstorm (A). The girdling root severely restricted wood development in the trunk (B).

Figure 5.130. Container-grown plants often develop circling roots. Here, circling roots indicate the size of the container at one point, while new roots have developed laterally from the root ball.

Symptoms

Symptoms of girdling or kinked roots may occur in young nursery plants or in mature trees. Trees with severe defects show symptoms at planting time and usually do not survive to maturity. If a tree has a kinked root of 90° or more and less that 20 percent of the root ball originates above the kink, the tree will typically fall over at the soil line if the supporting stake is removed (fig. 5.132).

In young trees with circling roots, the trunk may tend to move independently of the root ball, and the trees may fall over at the soil line when the stake is removed.

Girdling reduces the caliper growth (diameter) of roots and stems. Trunk tissue becomes compressed, and a swelling may occur just above the compressed area. Phloem transport is slowed, the entire plant becomes weakened, and vigor gradually declines. In the fall, affected trees tend to change color and defoliate earlier than other trees.

Figure 5.131. This girdling root on American elm *(Ulmus americana)* will limit trunk development and compromise its structural strength.

Figure 5.132. Young trees with girdling or kinked roots are prone to falling over, at the soil line when stakes are removed.

Figure 5.133. Girdling roots often develop at or near the soil surface. A root crown inspection can be conducted to determine whether girdling roots are present below ground.

Causes

Kinking and girdling typically result from poor nursery practices: either fast-growing species are not transplanted early enough or the root system of transplants is not properly placed in the new container. Each time a plant is transplanted into a larger container, it may develop girdling or kinked roots. Girdling and kinking are most detrimental to trees when they develop near the stem or in the center of the root ball. These defects cannot be corrected at planting since they affect the primary roots and correction would weaken the tree. Girdling roots that develop around the periphery of root ball are not usually a problem and can be corrected at planting.

Girdling roots may naturally develop in the landscape, typically 3 to 15 years after planting. If roots are cut during transplanting they may produce laterals at 90° angles from main roots that develop close to the trunk. As they increase in size, the laterals constrict the trunk above the girdling root. Trees with wet and shaded bark at ground level and trees planted in compacted soil tend to develop girdling roots. Compacted, glazed planting holes may encourage girdling roots as well.

Look-Alike Disorders

Symptoms of girdling and kinked roots include slow growth, poor color, premature leaf fall, and in some cases poor anchorage. These symptoms are caused by girdling- or kinking-induced aeration deficit, water deficit, and root disease.

Diagnosis

The root zone must be inspected to diagnose a suspected kinked or girdling root disorder. Girdling roots are typically on or near the soil surface (fig. 5.133). Carefully dig around the root crown and look for a root encircling the trunk and an indentation or reduction in trunk diameter. If the tree has a basal flare, girdling roots are not likely to be present. If possible, examine main roots radiating from the trunk for kinks or girdling roots.

Sensitive and Tolerant Species

Fast-growing trees and tap-rooted species are prone to girdling and kinking. Although an extensive characterization of tree species has not been conducted, elm (*Ulmus* spp.), Norway and sugar maple (*Acer* spp.), oak (*Quercus* spp.), and pine (*Pinus* spp.) have been observed to develop girdling roots.

Figure 5.134. In container stock, circling or kinked roots on the outside of the root ball should be cut or straightened before planting.

Remedies

Prevention is primary. Inspect trees prior to planting for root defects. Do not plant trees with severely kinked or girdling roots. Prevent soil from collecting around the trunk base to minimize the development of adventitious roots. Inspect the roots at the collar and remove roots that become a problem.

The location and severity of girdling roots determine whether and how much the injury can be corrected. If the circling roots at the point where the trunk meets the ground or at the center near the stem encircle more than 80 percent of the root system, the tree is likely to grow slowly and may eventually die. If the roots are circling at the periphery of the root ball, this may be corrected at planting time (fig. 5.134).

Girdling roots should be cut where they attach to the trunk or to another root and then again beyond the point of the girdle. If the roots have grafted to the trunk they should be left in place. If they have not, cut a V-shaped notch halfway through the root midway along the girdling section. The notched root will still supply the top of the tree but as the trunk expands, the root will break at the notch and can be removed.

Mechanical Injury

Mechanical injury includes damage from rubbing, cutting, shredding, constricting, crushing, or puncturing. Many things cause it, including hardware (stakes, ties, ropes, bolts, props, protective cages, injection apparatus) (fig. 5.135); equipment (trenchers, string trimmers, lawn mowers, saws, garden tools); vehicles (cars, trucks, tractors, backhoes, caterpillars, loaders); humans (vandalism); and miscellaneous agents (high-pressure irrigation spray, golf balls, tree spikes, carving or engraving) (fig. 5.136). Mechanical injury caused by animals (squirrels, deer, woodpeckers, sapsuckers, dogs, gophers, cows) is considered to be vertebrate pest damage, a biotic disorder.

Figure 5.135. Staking hardware often causes mechanical injury to young trees. Here, a tie was installed too tightly and left on too long (A). Eye bolts can cause substantial mechanical injury (B).

Symptoms

Mechanical injury may range in severity from minor wounds to death of the whole plant. All plant parts can be affected. Typically, bark and underlying tissues are damaged on trunks and branches; a section of bark may be removed, stripped, or severely constricted. If damage encircles or girdles the stem, water, nutrient, and photosynthate transport are interrupted (fig. 5.137). Mechanical damage or removal of roots can severely affect both vascular transport and structural stability. Often, mechanically injured plants are prone to further injury from bark beetles, wood-boring insects, and wood decay or canker fungi. Plant condition, size, and time of year influence the potential for recovery from mechanical injury.

Symptoms vary depending on the time, location, and degree of injury. If damage occurs in spring or early summer, the bark will "slip" or be easily knocked off. If large sections of bark are removed, damaged, or constricted, symptoms include wilting, stunting, premature leaf fall, and foliar discoloration. Small trees can be severely damaged in this manner by string trimmers and lawn mowers (fig. 5.138).

Girdling hardware (ropes, wires, ties, etc.) may cause a differential enlargement of the stem diameter: larger above the girdle and smaller below (fig. 5.139). Canopy symptoms include chlorosis, slow growth, and leaf drop (canopy thinning). If the girdle is severe, the trunk or branch may break.

Figure 5.136. Mechanical injury can occur in many ways. Here, golf balls have caused considerable injury to this tree (A). Bark injury can occur when high-pressure irrigation heads are directed at trees (B).

Figure 5.139. Ties left on trees can restrict trunk development. Here, trunk diameter growth of Monterey pine *(Pinus radiata)* was greater above the tie than below it. This form of mechanical injury reduces structural strength and increases the potential for failure.

Figure 5.137. The metal collar placed around this London plane tree *(Platanus × acerifolia)* will limit trunk growth, interfere with canopy development, and reduce structural stability.

Figure 5.138. If mechanical injury encircles the entire trunk or stem, transport of water and minerals is interrupted and vigor declines. Here, a string trimmer removed most of the bark at the trunk base.

Figure 5.140. Mechanical injury to roots can be lethal. These coast redwood *(Sequoia sempervirens)* were killed by root cutting during construction of a nearby house and perimeter wall (A). During sewer line replacement, over 75 percent of the roots were cut on this southern magnolia *(Magnolia grandiflora)*, resulting in extensive leaf drop and branch dieback.

Figure 5.141. Root injury from construction activities may be difficult to diagnose. Impacts may have occurred several years before symptoms appear. These valley oak *(Quercus lobata)* declined in the years following construction of adjacent buildings and a parking lot.

Mechanical injury to roots can severely harm plants (fig. 5.140). Root cutting (e.g., trenching) may cause a tree or shrub to decline rapidly and die. Plants that are injured but not killed often exhibit poor vigor, premature leaf drop, and injury by secondary pests. In many cases, injury or death may not be apparent in the above-ground portion of the plant for several years after the injury has occurred (fig. 5.141). If large-diameter roots are cut, structural support (anchorage) will be compromised (fig. 5.142).

Grade changes, particularly when the grade is lowered, may also injure roots because many roots may be cut in the process. Symptoms include poor vigor and general decline. Decline due to root cutting may be difficult to diagnose because symptoms may not be apparent right away and damaged roots are not easily observed unless the area is excavated.

Vandalism (intentional damage to plant parts) is common in many urban areas and can come in many forms. Typically, branches and stems are broken (fig. 5.143). In some cases, a section of trunk bark is removed, girdling the tree (fig. 5.144). Young trees are most severely affected.

Look-Alike Disorders

Mechanical injury may produce symptoms similar to those caused by girdling and kinked roots, water deficit, aeration deficit,

Figure 5.142. Root loss can reduce anchorage and tree stability. This Monterey pine *(Pinus radiata)* uprooted during a winter storm after roots had been cut.

Figure 5.143. Vandalism can be considered to be intentional mechanical damage by humans. Here, a young southern magnolia *(Magnolia grandiflora)* was vandalized and needs to be replaced.

root and canker disease, bark beetle injury, wind damage, and nutrient deficiency.

Diagnosis

In most cases, mechanical injury is not difficult to diagnose. Physical injury to the trunk and branches is usually noticeable.

- Inspect the bark for injury (removal, shredding, rubbing, or stripping).
- If hardware is attached (ties, ropes, bolts, etc.), assess damage to the bark and underlying tissue. Is hardware embedded in the bark and wood? Check whether the stem diameter is larger above the hardware.
- Consider the plant location and determine whether vehicle damage was possible.
- Learn the history of the site and determine the type of equipment used around the damaged plant. Have fertilizer or pesticide injections been made? Are trees protected from mowers or string trimmers? Is the irrigation system causing injury?
- Mechanical injury to roots is difficult to diagnose, especially if the injury occurred months or years before symptoms were observed. This type of injury is often associated with construction and development. Excavation is usually necessary to confirm mechanical injury to roots. Pneumatic excavation tools are useful for this purpose.

Sensitive and Tolerant Species

Although all species of trees and shrubs are susceptible to mechanical injury, some species are more tolerant than others. Trees with thick, spongy bark, such as cork oak *(Quercus suber)*, are more tolerant than trees with smooth, thin bark are more easily damaged.

Remedies

If ties, ropes, wires, or other hardware have girdled the stem, remove the material if possible. If hardware has become embedded, cut or slit it in several places around the stem so the plant can overgrow the obstruction. If the tops of trees have been broken, new growth may develop below the injury;

prune out the broken section. Restructuring of scaffold limbs or selection of a new central leader may be needed. Protect trees from sources of mechanical injury.

- Install protective staking.
- Remove support stakes and hardware when no longer needed.
- Maintain a vegetation-free area around each tree or shrub.
- If string trimmers are used, install a protective sleeve around the tree base. Plant trees where cars or other vehicles cannot accidentally damage them.
- Protect trees with barrier fencing during construction.
- Use protective screens where vandalism may occur.
- Avoid regular trunk or root crown injections, especially for species prone to bacterial wetwood, such as elm (*Ulmus* spp.), hemlock (*Tsuga* spp.), magnolia (*Magnolia* spp.), maple (*Acer* spp.), mulberry (*Morus* spp.), oak (*Quercus* spp.), poplar (*Populus* spp.), tulip tree (*Liriodendron tulipifera*), and species with thin bark.
- Avoid using tree spikes when climbing trees, especially on thin-barked trees.
- Avoid trenching across roots. If trenching must be done, limit root damage by trenching radially from the trunk rather than perpendicularly to the trunk. Start from outside the dripline and move in until roots greater than 1 inch (2.5 cm) in diameter are encountered.
- Stake or guy young trees properly and inspect frequently to avoid injury from ties or other hardware.
- Shorten stakes to avoid rubbing injury.

Figure 5.144. Vandalism can be lethal. This white mulberry *(Morus alba)* was killed by a vandal who girdled the trunk (A). Removing a section of bark from around a tree trunk (intentionally or not) is often fatal (B).

Graft Incompatibility

Graft union abnormalities include the failure of the stock (root) and scion (top) tissues to unite during grafting and the partial union of stock and scion tissues, a condition referred to as delayed, or partial, incompatibility.

In some cases of partial incompatibility, the stock and scion are united by callus tissue but not by conducting tissues, causing the plant to be short-lived. In other cases, the stock and scion are united by xylem tissues but not by phloem tissues, causing the stock to slowly starve and die. In still other cases, no interlocking growth of stock and scion xylem tissues occurs, causing the plant to have a weak union that is subject to breakage.

Symptoms

A plant dies if the stock and scion fail to unite at grafting. Partial graft incompatibility has several symptoms. The growth of the scion during spring may be erratic when the stock and scion grow at different rates or begin to grow at different times (fig. 5.145). Plants may also suffer various symptoms of physiological stress as a result of partial graft incompatibility, including stunting, chlorosis, premature fall leaf color and leaf drop, and a general decline in vigor (fig. 5.146). Plants may also produce heavier than normal amounts of flowers and fruit. There may be an obvious swelling or overgrowth of woody tissues above, below, or at the graft union, and new growth may sprout from the rootstock (fig. 5.147). In some cases, the scion (or the entire plant) eventually dies as a result of the accumulated stresses. Plants suffering from partial graft incompatibility may also break at the graft union.

Figure 5.145. Partial graft incompatibility (overgrowth) of Modesto ash (*Fraxinus velutina* var. *glabra* 'Modesto') budded on green ash *(Fraxinus pennsylvanica* var. *lanceolata)* rootstock.

Figure 5.146. Partial graft incompatibility caused the decline of this Yarwood sycamore (*Platanus acerifolia* 'Yarwood').

Occurrence

Partial graft incompatibilities are relatively uncommon, but they sometimes occur in budded landscape trees. Examples are 'Yarwood' sycamore (*Platanus acerifolia* 'Yarwood') budded to London plane (*Platanus acerifolia*) rootstock, and Modesto ash (*Fraxinus velutina* var. *glabra* 'Modesto') budded to green ash (*Fraxinus pennsylvanica* var. *lanceolata*) rootstock.

Look-Alike Disorders

Anything that damages the root system, including girdling roots, diseases, root pruning, soil compaction, adverse soil temperatures, aeration deficit, and poor drainage may produce symptoms resembling partial graft incompatibility. Mechanical girdling of the trunk may also cause similar symptoms.

Diagnosis

Begin by confirming that the tree was budded or grafted. Also, look for stunted growth and early fall color. Partial graft incompatibility is often easy to diagnose from the obvious woody overgrowth of the scion on the stock.

Remedies

A valuable tree affected by graft incompatibility may be saved by in-arch grafting (see Hartman and Kester 1975). In most cases the tree should be removed.

Figure 5.147. Rootstock sprouting caused by partial graft incompatibility in Yarwood sycamore (*Platanus acerifolia* 'Yarwood').

Herbicide Phytotoxicity

Herbicides that are used to control weeds in or near landscapes sometimes injure landscape plants. The injury is usually caused by a herbicide not being used in accordance with label instructions, such as using it on or around sensitive nontarget plants, increasing the rate of application, or by carelessness (e.g., spraying during windy conditions).

Often, the identity of the herbicide or the herbicide family can be determined by the injury symptoms. Symptoms can be divided into five main categories: root and shoot stunting, leaf chlorosis, leaf necrosis, leaf spotting, and leaf and shoot malformations.

Diagnosing Herbicide Phytotoxicity

When diagnosing herbicide phytotoxicity, it is important to know how the injured plant should normally appear: for example, certain herbicides produce leaf symptoms resembling variegation common to some landscape plant species. Also, careful observation of surrounding plants, including weeds, may help establish where the herbicide was applied and whether the injury resulted from aerial drift or root absorption: for instance, necrotic leaf spots, especially on leaves of the same age, may indicate an aerial drift of a postemergence contact herbicide. Interveinal chlorosis and marginal necrosis, especially on older leaves on one side of a tree, may indicate injury caused by a preemergence systemic herbicide. Symptoms often begin on the same side of the tree where a preemergence herbicide was applied.

Learn what herbicide may have been used in the vicinity of plants exhibiting phytotoxicity symptoms. For example, find out whether herbicides were used in turfgrass to control broadleaf weeds or on a nearby driveway to prevent weed growth, and if so, which ones. A recent history of herbicide use may confirm suspicions created by the observation of symptoms.

Also, be familiar with the mode of action of the herbicide used or suspected. This helps in making an accurate diagnosis and may be necessary for determining what steps can be taken to remedy the injury. However, it is often difficult to learn the identity of the herbicide causing phytotoxicity, especially when newly planted trees are injured by long-lived soil-residual herbicides, or when plants are injured by a herbicide spray drifting from a distance, or when the injury is caused by carelessness.

Soil and leaf analyses may be used to help diagnose herbicide injury. A soil analysis is useful only for preemergence (soil-applied) herbicides. A leaf analysis may be useful for detecting systemic herbicides or herbicides that have contacted leaf surfaces, as in drift injury. Samples should be taken as quickly as possible following injury, as herbicides may quickly degrade in soil and on leaf surfaces, reducing the detectable concentration. It is usually necessary to provide the laboratory with the name of the suspected herbicide.

A herbicide may exist in the soil at a concentration sufficient to affect plant growth but be below the minimum detection limit of a laboratory. The estimated concentration in soil may be calculated if time of application, rate, and herbicide half-life are known. This estimate may be compared with the minimum detection limit to judge whether laboratory analyses are likely to be informative.

Some preemergence herbicides used for total vegetation control may persist in the soil for many months, especially when applied at high rates. In cases where plants have been injured by such herbicides, symptoms may also persist for many months. Although plant symptoms remain, the soil-active concentration may drop below a level necessary to cause injury. Differences in soil texture and absorptive qualities greatly influence herbicide activity. A herbicide level of 1 ppm in sandy soils is more active than 1 ppm in clay soils.

When diagnosing herbicide injury from soil-applied herbicides, it is often helpful to perform a bioassay using test plants. Depending on the herbicide and its concentration, the test plants may be stunted, distorted, or killed. A bioassay is an inexpensive technique for determining whether the plant symptoms are caused by a herbicide or something else. Use the following procedure to conduct a bioassay.

- Take soil from depths of 0 to 2 inches, 2 to 4 inches, and 4 to 6 inches (0 to 5 cm, 5 to 10 cm, and 10 to 15 cm) in suspected treated and known untreated areas. Collect 2 or 3 containers (paper cups or small plastic flower pots) of soil from each area. A greater depth may be necessary if you suspect an extremely soluble herbicide.
- Plant seeds of sensitive and tolerant species in the treated and untreated soils (see table 5.27 for indicator species). The plants that are sensitive to herbicides will show phytotoxicity symptoms or be killed. The tolerant plants will be damaged less or remain uninjured. Use more than one sensitive plant species to help determine which herbicide may be present.
- Grow the plants under good cultural conditions for at least 20 days. Be careful not to over- or underwater them.

TABLE 5.27. Indicator plants for herbicide families

Herbicide family	Chemical name	Trade name	Tolerant plants	Susceptible plants
arylurea	diuron tebuthiuron	Karmex Spike	groundsel *(Senecio vulgaris)*, oat *(Avena sativa)*	sugar beet *(Beta vulgaris)*, tomato *(Lycopersicon esculentum)*
benzoic acid	dicamba	Banvel	milo *(Sorghum vulgare)*, ryegrass (*Festuca* spp.)	alfalfa *(Medicago sativa)*, bean *(Phaseolus vulgaris)*, safflower *(Carthamus tinctorius)*
dinitroaniline	benefin oryzalin pendimethalin prodiamine	Balan Surflan Pendulum, Pre-M, Prowl Barricade, Endurance	carrot *(Daucus carota)*, cotton *(Gossypium hirsutum)*, safflower *(Carthamus tinctorius)*, sunflower *(Helianthus annuus)*, tomato *(Lycopersicon sculentum)*	barnyardgrass *(Echinochloa crus-galli)*, millet *(Panicum miliaceum)*, milo *(Sorghum vulgare)*, oat *(Avena sativa)*, sugar beet *(Beta vulgaris saccharifera)*
nitrile	dichlobenil	Casoron, Dyclomec	corn *(Zea mays)*, milo *(Sorghum vulgare)*	carrot *(Daucus carota)*
pyridazine	norflurazon	Solicam, Predict	mint (*Mentha* spp.), pigweed (*Amaranthus* spp.)	oat *(Avena sativa)*, barley *(Hordeum vulgare)*, bean (*Phaseolus vulgaris)*, corn *(Zea mays)*
thiocarbamate	EPTC	Eptam, others	alfalfa *(Medicago sativa)*, bean *(Phaseolus vulgaris)*	barley *(Hordeum vulgare)*, milo *(Sorghum vulgare)*
triazines	atrazine simazine	Aatrex, others Princep, others	corn *(Zea mays)*, crabgrass (*Digitaria* spp.), milo *(Sorghum vulgare)*	lettuce *(Lactuca sativa)*, sugar beet *(Beta vulgaris saccharifera)*, tomato *(Lycopersicon esculentum)*
uracil	bromacil	Hyvar	citrus (*Citrus* spp.), mint (*Mentha* spp.)	all plants except citrus (*Citrus* spp.) and mint (*Mentha* spp.)

Source: Lange, Elmore, and Saghir 1969.

- Evaluate tolerant and sensitive plant species, comparing the plant growth and appearance in suspected treated soil with plants grown in known untreated soil. Check for stunting, yellowing or other discoloration, distortion of stems, and leaves and root swelling and stunting.

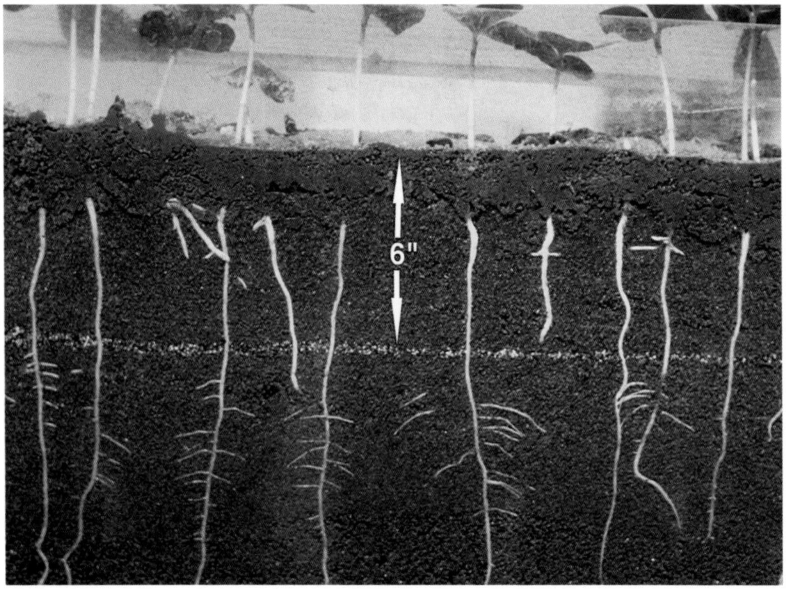

Figure 5.148. Lateral root growth in cotton *(Gossypium hirsutum)* seedlings is suppressed in the top 6 inches (15 cm) of trifluralin-treated soil. Lateral roots grow normally in untreated soil (below).

Figure 5.149. Lateral roots of cotton *(Gossypium hirsutum)* seedlings growing normally in the top 6 inches (15 cm) of soil, but suppressed in trifluralin-treated soil (below).

Root and Shoot Stunting and Distortion

Symptoms

Certain preemergence herbicides may cause stunting and distortion of roots and shoots. Injured roots are short, thickened, and swollen at the tips. The growth of secondary or lateral roots is suppressed or delayed (figs. 5.148–149). In seedlings, herbicides such as napropamide cause roots and shoots to be stunted, but the root tips do not enlarge. Herbicides such as oryzalin, isoxaben, and trifluralin cause root tips to enlarge into globose structures and shoots to be stunted; lateral roots fail to develop, and in some plant species, the leaves are also deformed. In seedlings, the stem may swell at or above the soil line. Injured plants may appear normal but may wilt more readily as the stunted root system fails to absorb the necessary moisture. Even in some established woody plants, certain preemergence herbicides can cause stems to enlarge and become brittle; examples include oryzalin in Monterey pine *(Pinus radiata)*, and prodiamine in true firs (*Abies* spp.).

Occurrence

The preemergence herbicides that can produce root and shoot stunting and distortion include (see table 5.28)

- benefin
- dithiopyr
- isoxaben (stunts broadleaf plants only, not grasses)
- oryzalin
- pendimethalin
- prodiamine
- trifluralin (affects grasses more than broadleaf plants)

Injury may occur from herbicide overapplication to seedlings or to poorly rooted young plants or young nursery stock, or from incorporation too deeply into the root zone. Established plants with well-developed root systems can typically tolerate even high rates of these herbicides.

Groundcover liners planted in soil treated with a preemergence herbicide will be injured. This is especially true of large leaf iceplant (*Carpobrotus* spp.), which is planted unrooted; rooting is inhibited for approximately 3 months until the herbicide degrades. In most cases root stunting occurs primarily in the top 1 to 2 inches (2.5 to 5 cm) of soil.

TABLE 5.28. Herbicides that can cause root and shoot stunting and distortion

Herbicide family	Chemical name	Trade name
	Preemergence	
amide	metolachlor napropamide	Pennant Devrinol
benzamide	isoxaben	Gallery
dinitroanline	benefin oryzalin oryzalin + oxyfluorfen pendimethalin prodiamine trifluralin	Balan Surflan Rout Pendulum, Pre-M, Prowl Barricade, Endurance Treflan, others
pyridine	dithiopyr	Dimension
thiocarbamate	EPTC DCPA	Eptam, others Dachtal
	Postemergence	
sulfonylurea	halosulfuron	Manage

Figure 5.150. Interveinal chlorosis in sweetgum *(Liquidambar styraciflua)* foliage caused by simazine.

Look-Alike Disorders

Plant parasitic nematodes can stunt root systems and cause swelling of roots, and soilborne diseases can restrict root development. Aluminum toxicity can also cause root stunting. Soil compaction reduces root development but does not prevent the development of root hairs or lateral roots.

Remedies

Activated charcoal (carbon) may be incorporated into soil to adsorb and remove certain preemergence herbicides. Organic matter, such as compost or manure, may also adsorb certain preemergence herbicides. Improve the vigor of herbicide-injured plants with good culture, especially proper irrigation practices. Do not use excessive irrigation or fertilizers to invigorate weakened plants. Soluble herbicides may be removed from the soil by leaching, but off-site movement and increased plant injury are possible following water application.

Leaf Chlorosis: Marginal, Veinal, Interveinal, Entire, or Mottled

Symptoms

Certain preemergence systemic herbicides may inhibit photosynthesis, causing various patterns of chlorosis, depending on the type and concentration of the herbicide (fig. 5.150). The herbicides that cause these symptoms are absorbed by the root system, then move through xylem tissues to the leaves. In most cases, symptoms occur on mature foliage first, since older leaves have had more time to accumulate the herbicide. In some cases, however, as with norflurazon, the symptoms occur first in young foliage. Symptoms are usually first seen and are more severe on the side of the plant where the herbicide was placed. Chlorotic leaf margins and interveinal areas often become progressively necrotic (fig. 5.151).

With some preemergence herbicides (e.g., norflurazon), the symptoms are a veinal yellow to whitish chlorosis (vein clearing), with a striking contrast between green and chlorotic areas of the leaf (figs. 5.152–153).

Figure 5.151. Interveinal chlorosis and necrosis in fig *(Ficus carica)* foliage caused by diuron.

Figure 5.152. Veinal chlorosis in sawleaf zelkova *(Zelkova serrata)* foliage (upper shoot) and overall chlorosis in xylosma *(Xylosma congestum)* foliage (lower shoot). Both symptoms were caused by diuron-bromacil combination.

Figure 5.153. Veinal chlorosis in almond *(Prunus dulcis)* foliage caused by norflurazon.

Occurrence

Photosynthesis-inhibiting preemergence herbicides that can produce marginal, veinal, interveinal, entire, or mottled chlorosis in leaves include (table 5.29; fig. 5.154)

- bromacil
- dichlobenil
- diuron
- prometon
- simazine
- tebuthiuron

Spring and summer applications of glyphosate (Roundup), a postemergence systemic herbicide, can cause overall chlorosis of leaves (fig. 5.155).

Injury is usually found when preemergence systemic herbicides are used for total vegetation control within the root zone of desired plants. When used at high rates, the symptoms in desirable woody plants may persist for more than a year. Trees connected by root grafts may be injured when a postemergence systemic herbicide translocates from tree to tree ("flashback") (fig. 5.156). Flashback injury is not known to occur with soil-applied herbicides. Injury from glyphosate usually occurs as the result of spray drifting onto leaves during spring or summer.

Herbicide Persistence in the Environment

Like other chemicals, herbicides vary greatly in persistence. All are degraded in the environment by chemical and microbial processes. Herbicide concentration typically follows an exponential decay curve, the steepness or slope of which depends on the half-life of the compound in the environment. Most herbicides break down in a few weeks; others may remain in the soil for several months, and a few may last for more than a year. Because damaged leaves are not repaired by plants, the presence of foliar symptoms does not necessarily imply the continued presence of the herbicide in the environment.

Look-Alike Disorders

Interveinal chlorosis symptoms may be confused with micronutrient (e.g., iron, manganese) deficiency symptoms. However, micronutrient deficiency symptoms generally occur first in new foliage, while injury caused by many root-absorbed preemergence herbicides usually occurs first in older foliage (an exception is damage caused by norflurazon) (figs. 5.157–158).

Figure 5.154. Interveinal necrosis in London plane *(Platanus × acerifolia)* foliage caused by tebuthiuron. Note that new growth is initially unaffected.

TABLE 5.29. Herbicides that can cause marginal, veinal, interveinal, entire, or mottled leaf chlorosis

Herbicide family	Chemical name	Trade name
	Preemergence	
arylurea	diuron tebuthiuron	Karmex Spike
arylurea + uracil	diuron + bromacil	Krovar
triazine	prometon simazine	Pramitol Princep, others
imidazolinone	imazapyr imazaquin	Arsenal Image
nitrile or misc.	dichlobenil	Casoron, Dyclomec
pyridazine	norflurazon	Solicam, Predict
sulfonylurea	sulfometuron	Oust
uracil	bromacil	Hyvar
	Postemergence	
organophosphate	glyphosate	Roundup
sulfonylurea	halosulfuron	Manage

Also, marginal leaf burn begins earlier in the development of herbicide injury symptoms than in chlorosis symptoms caused by micronutrient deficiency. The natural yellow leaf variegations of some plant species such as true myrtle (*Myrtus* spp.) and euonymus (*Euonymus* spp.) may resemble chlorosis patterns caused by preemergence herbicide injury. Blotchy chlorosis caused by herbicides may be confused with symptoms of certain virus diseases, such as mosaic in roses. Blotchy chlorosis may also be caused by sucking insects such as lace bugs, and chronic drought and high soil salinity can also cause leaf chlorosis.

Remedies

Incorporate activated charcoal (carbon) into the soil to adsorb and remove certain preemergence herbicides. Organic matter, such as compost or manure, may also adsorb certain preemergence herbicides. Improve the vigor of herbicide-injured plants with good culture, especially proper irrigation practices. Do not use excessive irrigation or fertilizers to invigorate weakened plants. Soluble herbicides may be removed from the soil by leaching, but off-site movement and increased plant injury are possible following water application.

Leaf Necrosis: Marginal, Interveinal, Blotchy, or Entire

Symptoms

At high rates, preemergence herbicides may cause necrosis in leaves of all ages before any chlorosis patterns can develop (figs. 5.159–160). High concentrations of postemergence contact herbicides may cause overall leaf necrosis or marginal leaf necrosis without chlorosis or spotting (table 5.30).

Occurrence

Preemergence systemic herbicide injury usually occurs when herbicides such as bromacil, dichlobenil, diuron, prometon, simazine, or tebuthiuron are used for total vegetation control within the root zone of

Figure 5.155. Overall chlorosis of escallonia *(Escallonia rubra)* foliage caused by spring application of glyphosate.

Figure 5.156. Chlorosis in valley oak *(Quercus lobata)* foliage caused when glyphosate moved from treated to untreated tree via root grafts.

Figure 5.157. Other abiotic disorders can cause symptoms resembling herbicide phytotoxicity, such as iron deficiency, seen here in sweetgum *(Liquidambar styraciflua)*. Note that the most severe symptoms occur on new leaves.

desired plants. When used at high rates, these materials may cause symptoms that persist for more than a year. Leaf necrosis from postemergence herbicides such as diquat and paraquat occurs as a result of drift of spray onto leaves. Injury may also occur when herbicide-contaminated spray equipment is used to treat plants with insecticides or fungicides.

Look-Alike Disorders

Leaf necrosis symptoms caused by herbicides may be confused with symptoms caused by high concentrations of soil salts, leaf spot diseases, drought, or freezing temperatures. Foliar applications of fertilizers, pesticides, or other chemicals such as plant growth regulators may also cause phytotoxicity, resulting in leaf necrosis.

Remedies

Incorporate activated charcoal (carbon) into soil to adsorb certain preemergence herbicides. Improve the vigor of herbicide-injured plants with good culture, especially proper irrigation practices. Moist soils favor the soil microorganisms that degrade chemicals.

Chlorotic or Necrotic Leaf Spots

Symptoms

The aerial drift of certain postemergence contact herbicides may cause chlorotic and necrotic spots on leaves. The symptoms occur only on the leaves contacted by the chemical and not on leaves that grow later (figs. 5.161–162). In a group of plants, leaf spotting would start on one side of the plant, then diminish gradually and uniformly away from the initial contact. The amount of chemical contacting the leaves determines the extent of the injury. Chlorotic spots may become necrotic with time.

Figure 5.158. Atrazine injury in toyon *(Heteromeles arbutifolia).* Note that the most severe injury is on oldest leaves.

Figure 5.159. Necrosis in Canary Island pine *(Pinus canariensis)* needles caused by high rate of tebuthiuron.

Figure 5.160. Necrosis in Modesto ash (*Fraxinus velutina* var. *glabra* 'Modesto') foliage caused by prometon.

Occurrence

Herbicides that produce chlorotic spots on leaves include diquat, oxyfluorfen, paraquat, and pelargonic acid, all postemergence contact herbicides. Injury from these herbicides usually occurs as a result of aerial drift. Leaf spots can also result from systemic postemergence herbicides such as sethoxydim (Poast) and fluazifop (Fusilade), used to selectively control grass weeds in ornamentals. Injury may also occur when herbicide-contaminated spray equipment is used to treat plants with insecticides or fungicides.

Certain preemergence herbicides may also cause spotting of foliage (table 5.31). When granular herbicides such as oxadiazon (Ronstar) remain on leaves following an application, chlorotic spots approximately the size of the granules often develop on the leaves. The spots eventually become necrotic or "burn" (fig. 5.163).

Look-Alike Disorders

Other chemicals, such as insecticides, fungicides, and plant growth regulating chemicals may also cause chlorotic or necrotic leaf spotting. Foliar applications of fertilizers may cause leaf spotting as well. Certain fungal leaf spot diseases, such as Entomosporium leaf spot (caused by *Entomosporium maculatum*) and shot hole disease *(Wilsonomyces carpophilus),* also produce leaf spot symptoms that may be confused with herbicide-caused leaf spotting. However, necrotic spots do not drop out of herbicide-spotted leaves as with shot hole disease. Also, under the proper environmental conditions, leaf-spotting diseases continue to infect new leaves, while spots caused by herbicides occur only on leaves present at the time the drift took place (fig. 5.164).

Remedies

Improve the vigor of herbicide-injured plants with good culture, especially proper irrigation practices.

TABLE 5.30. Herbicides that can cause marginal, interveinal, entire, or blotchy leaf necrosis without chlorosis or spotting

Herbicide family	Chemical name	Trade name
Preemergence		
arylurea	diuron tebuthiuron	Karmex Spike
arylurea + uracil	diuron + bromacil	Krovar
imidazolinone	imazapyr	Arsenal
nitrile or misc.	dichlobenil	Casoron, Dyclomec
pyridazine	norflurazon	Solicam, Predict
sulfonylurea	sulfometuron	Oust
triazine	atrazine prometon simazine	Aatrex, others Pramitol Princep, others
uracil	bromacil	Hyvar
Postemergence		
bipyridinium	diquat paraquat	Ortho Diquat, Reward Gramoxone Super
fatty acids	pelargonic acid	Scythe
Preemergence and postemergence		
diphenyl-ether	oxyfluorfen	Goal

TABLE 5.31. Herbicides that can cause chlorotic or necrotic leaf spots

Herbicide family	Chemical name	Trade name
Preemergence		
dinitroanaline +diphenyl-ether	oryzalin + oxyfluorfen	Rout
miscellaneous	oxadiazon	Ronstar
Postemergence		
aryl triazinone	carfentrazone	Shark
bipyridinium	diquat	Ortho Diquat, Reward
cyclohexenone	sethoxydim paraquat	Poast Gramoxone Super
fatty acids	pelargonic acid	Scythe
organic arsonate	MSMA	Bueno 6
oxyphenoxy propionate	fluazifop	Fusilade
miscellaneous	bentazon glufosinate	Basagran Finale
Preemergence and postemergence		
diphenyl-ether	oxyfluorfen	Goal

Malformed or Distorted Leaves and Shoots

Symptoms

The aerial drift or root absorption of certain selective translocated herbicides may cause "hormonelike" symptoms (table 5.32), including leaf cupping (leaf margins turning up or down), petiole twisting, downward bending of petioles, distorted and stunted fan-shaped leaves, prominent leaf veins, reduction of leaf intervein areas, calluslike growth on young woody stems, splitting or cracking of bark, delayed leafing in spring, and chlorosis. Because the herbicides that cause these symptoms are translocated, symptoms are generally distributed on new growth over the entire plant (figs. 5.165–166).

The aerial drift of glyphosate, a non-selective translocated herbicide, may cause delayed leafing in spring, shortened internodes, small leaves, leaf distortion and chlorosis (figs. 5.167–168). Translocated herbicides may also move from one plant to another via root grafts.

Occurrence

Selective herbicides, such as 2,4-D, dicamba, and triclopyr, used for broadleaf control in lawns may be absorbed by the roots of trees and shrubs growing nearby. Or, spray applications of the herbicides outside the landscape may drift to nearby desirable plants, causing symptoms of injury. Symptoms may persist in the plant the following year. Glyphosate spray applied for winter weed control may be absorbed through the thin or green bark of dormant woody ornamental plants such as roses. The injury becomes apparent the following spring. Injury may also occur when herbicide-contaminated spray equipment is used to treat plants with insecticides or fungicides.

Look-Alike Disorders

Shortened internodes and small leaves may be confused with zinc deficiency, an uncommon nutrient deficiency in landscape plants. Powdery mildew, a fungus disease, can cause distortion and witches' brooms of

Figure 5.161. Necrotic spots on bean (*Phaseolus* sp.) leaves caused by paraquat.

Figure 5.162. Necrotic spots on almond *(Prunus dulcis)* leaves caused by paraquat.

Figure 5.163. Necrotic spotting on lily (*Lilium* sp.) foliage caused by Ronstar granules catching in leaf bases.

Figure 5.164. Fungus leaf spot diseases, such as Entomosporium leaf spot (caused by *Entomosporium maculatum*), may be mistaken for herbicide phytotoxicity.

Figure 5.165. Malformation of hibiscus (*Hibiscus* sp.) shoot and leaves caused by 2,4-D.

Figure 5.166. Twisting and cupping of London plane *(Platanus × acerifolia)* leaves (right) caused by 2,4-D. Normal foliage on left.

Table 5.32. Herbicides that can cause malformed or distorted leaves and shoots

Herbicide family	Chemical name	Trade name
	Preemergence	
amide	metolachlor	Pennant
imidazolinone	imazapyr imazaquin	Arsenal Image
sulfonylurea	sulfometuron	Oust
	Postemergence	
benzoic acid	dicamba	Banvel
organophosphate	glyphosate	Roundup
phenoxy	2,4-D 2,4-DP (diclorprop) MCPP (mecoprop)	Several Several Several
pyridine carboxcylic acid	triclopyr	Garlon, Turflon
sulfonylurea	halosulfuron	Manage, Permit
miscellaneous	sulfasate	Touchdown

new growth in trees and shrubs (see fig. 3.18, p. 27). However, powdery mildew infections are readily diagnosed by the white to gray powdery growth on leaves and shoots. Dikegulac sodium (Atrimmec), a plant growth regulator, also causes similar symptoms.

Remedies

Improve the vigor of herbicide-injured plants with good culture, especially proper irrigation practices.

Figure 5.167. Injury to rose (*Rosa* sp.) caused by winter application of glyphosate.— Roundup

Figure 5.168. Injury to almond *(Prunus dulcis)* caused by late-summer to fall application of glyphosate.

Other Chemical Phytotoxicities

Insecticides, Fungicides, Plant Growth Regulators

Phytotoxicity may be caused by the misapplication of insecticides, fungicides, and plant growth regulating chemicals, or when environmental or cultural conditions predispose the plant to injury from these chemicals.

Symptoms

Chemical phytotoxicity symptoms include yellow to brown leaf spots; chlorosis or necrosis of leaf tips, leaf margins and interveinal areas; overall leaf chlorosis or necrosis; leaf curling; leaf stunting; and premature leaf drop. The injury symptoms do not spread with time or move to previously uninjured foliage or plants. Oil solvents or carriers in a pesticide (not the pesticide's active ingredient) may be responsible for producing the injury symptoms or making the symptoms more severe than normal.

Occurrence

Phytotoxicity from insecticides, fungicides, or plant growth regulators usually is caused by applying the chemicals at higher than recommended rates, especially when environmental conditions do not favor rapid drying of spray or when the plant is at a susceptible age. Chemical phytotoxicity can also be caused by treating plant species that are not listed on the product label, or when spray equipment contaminated with herbicides is used to apply insecticides, fungicides, or plant growth regulators.

Phytotoxicity is more likely to occur on water- or heat-stressed plants or plants that have low vigor. For example, olive trees suffering from drought stress may be partially defoliated by plant growth regulator sprays being used to eliminate nuisance fruits (fig. 5.169). Plants sprayed with horticultural oils during freezing weather may be seriously damaged. Spray oils may also remove the bluish cast from the foliage of glaucous

Figure 5.169. Injury to olive *(Olea europaea)* caused by an application of naphthaleneacetic acid (NAA) to control nuisance fruit.

Figure 5.170. These Chinese evergreen elm *(Ulmus parvifolia)* were treated with paclobutrazol (Clipper), a plant growth regulator used to control shoot growth. After 8 years, canopies have become very thin and some trees have died (A). Trees not under power lines (right) were not treated and do not show similar symptoms (B). Leaves on treated trees are closely spaced on branches (short internodes), darker green, and smaller than normal (C).

conifers such as blue spruce *(Picea pungens)*; affected trees usually recover rapidly, however. Phytotoxicity symptoms are difficult to diagnose without knowing the affected plant's cultural and chemical treatment history (fig. 5.170).

Look-Alike Disorders

Chemical phytotoxicity symptoms may be similar to symptoms caused by herbicides, excessive soil salinity, drought, and leaf spot diseases.

Diagnosis

Many of the same techniques used to diagnose herbicide phytotoxicity can be used to diagnose chemical phytotoxicity.

- Learn what chemicals may have been used to treat the plants showing phytotoxicity symptoms. A recent history of fungicide, insecticide, or plant growth regulator use may confirm suspicions created by the observation of symptoms.
- If the chemical can be identified, consult a label to determine if the injured plant is listed. Also determine what rates were used, as phytotoxicity is often caused by applications made at rates higher than label recommendations.
- Find out what the weather conditions were at time of chemical application. Phytotoxicity is more likely to occur during periods of high temperatures.

- Check the soil moisture. Certain chemicals, such as plant growth regulators used to eliminate nuisance fruit, may be more phytotoxic to water-stressed plants.
- Look for patterns of injury. Chemical phytotoxicity usually occurs only on leaves present when the pesticide or growth regulator was used. Subsequent growth will usually be free of the injury symptoms. Examine surrounding plants for similar injury; injury to a variety of plant genera helps confirm phytotoxicity.
- A leaf analysis may be useful for detecting chemicals that have contacted leaf surfaces. Samples should be taken as quickly as possible following injury, since chemicals degrade on leaf surfaces, affecting the detectable concentration. It is usually necessary to provide the laboratory with the name of the suspected chemical.

Remedies

Improve vigor of plants injured by phytotoxicity with good culture, especially proper irrigation practices.

Naturally Occurring Plant and Soil Toxins

Several organic compounds (toxins) may harm plant health under certain conditions. Toxic compounds may be excreted into the soil by the roots of certain plants or leached from the leaves of certain species by rain, inhibiting the growth of other plant species growing nearby. This inhibition is referred to as *allelopathy*. Toxins may also be produced by soil microbes as the microbes decompose plant residues above and below ground. The actual toxicity of these compounds and their longevity in the soil depends on numerous factors, including soil type, soil physical properties, soil moisture and oxygen content, the resistance of the compound to chemical and physical decomposition, and the plant species affected. Phytotoxic compounds that have been isolated from root excretions and soil and plant residues include

- organic acids: benzoic, *trans*-cinnamic, *p*-coumaric, cyanuric, *threo*-9, 10-dihydroxystearic, ferulic, *p*-hydroxybenzoic, 2-methyl-isonicotinic
- aldehydes: 3-acetyl-1-6-methoxybenzaldehyde, benzaldehyde, salicylaldehyde-vanillin
- amino acids: coumarin, esculentin, furanocoumarins, scopoletin
- glucosides: amygdalin, juglone, phlorizin

Symptoms

Organic toxins cause variable symptoms, depending on the plant species affected. Reported allelopathic effects include

- inhibition of germination of annual plant seeds
- inhibition of seedling growth
- darkened and swollen root tips
- distortion and stunting of roots
- inhibition of root and shoot growth in annual, perennial, and woody plants
- wilting, chlorosis, and premature fall colors in woody plants
- death of annual and woody plant seedlings and replants

Occurrence

There are relatively few documented cases in which organic toxins have been isolated, identified, and shown to have a growth-suppressing effect on plants grown in soil. Most of the reports of injury are to annual plants and involve inhibition of seed germination. In most cases, the studies were conducted in the laboratory, not in soil culture. Table 5.33 shows woody plants that have been reported to contain organic toxins and lists plants inhibited by the toxins in various studies; table 5.34 shows various annual plant species that are inhibited by fresh eucalyptus mulch; and table 5.35 shows plant species that exhibit growth inhibition from eucalyptus materials as ingredients of composts and potting mixes.

Look-Alike Disorders

Anything that damages roots may produce symptoms similar to those reported for organic toxins, including soil compaction, adverse soil temperatures, aeration deficits,

TABLE 5.33. Plants reported to contain organic toxins and plants affected

Plant containing organic toxins: Scientific name	Common name	Plant affected
Acer spp.	maple	wheat (*Triticum* spp.)
Artemisia absinthium	wormwood	fennel (*Foeniculum vulgare*), lovage (*Levisticum officinale*)
Castanea dentata	American chestnut	wheat (*Triticum* spp.)
Citrus spp.	citrus	citrus (*Citrus* spp.)
Cornus spp.	dogwood	wheat (*Triticum* spp.)
Cydonia oblonga	quince	quince (*Cydonia oblonga*)
Erica multiflora	heath	various annuals
Juglans cinerea	butternut	Swiss mountain pine (*Pinus mugo mughus*)
Juglans nigra	black walnut	alfalfa (*Medicago sativa*), apple (*Malus* spp.), asparagus (*Asparagus officinalis*), cherry (*Prunus* spp.), chrysanthemum (*Chrysanthemum* spp.), hydrangea (*Hydrangea macrophylla*), lilac (*Syringa vulgaris*), potato (*Solanum tuberosum*), sugar beet (*Beta vulgaris saccharifera*), tomato (*Lycopersicon esculentum*)
Liriodendron tulipifera	tulip tree	wheat (*Triticum* spp.)
Malus spp.	apple	apple (*Malus* spp.), pear (*Pyrus communis*)
Pinus spp.	pine	wheat (*Triticum* spp.)
Prunus armeniaca	apricot	cherry (*Prunus* spp.)
Prunus cerasifera	plum	pear (*Pyrus communis*)
Prunus persica	peach	peach (*Prunus persica*)
Prunus spp.	cherry	cherry (*Prunus* spp.), wheat (*Triticum* spp.)
Quercus spp.	oak	wheat (*Triticum* spp.)
Robinia pseudoacacia	black locust	barley (*Hordeum vulgare*)
Rosmarinus officinalis	rosemary	various annuals
Sorbus aucuparia	mountain ash	mountain ash (*Sorbus aucuparia*)

Source: Moje 1966.

TABLE 5.34. Plants exhibiting growth inhibition from fresh eucalyptus mulch

Eucalyptus species	Plants inhibited
E. citriodora	alyssum (*Alyssum* spp.)
E. ficifolia	amaranth (*Amaranthus* spp.)
E. globulus	barnyardgrass (*Echinochloa crus-galli*)
E. maculata	bermudagrass (*Cynodon dactylon*)
E. polyanthemos	bindweed (*Convolvulus arvensis*)
E. rudis	cornflower (*Centaurea cyanus*)
E. sideroxylon	large crabgrass (*Digitaria sanguinalis*) plantain (*Plantago* spp.) poppy (*Papaver* spp.)

Source: Downer 1993.

TABLE 5.35. Plants exhibiting growth inhibition from eucalyptus materials as ingredients of composts and potting mixes

Eucalyptus species	Plants inhibited
E. camaldulensis	bottlebrush (*Callistemon* spp.)
E. ficifolia	marigold (*Tagetes* spp.)
E. sideroxylon	petunia (*Petunia hybrida*)
E. viminalis	privet (*Ligustrum* spp.)
	radish (*Raphanus sativus*)

Source: Jarrell and Furuta n.d.

poor drainage, low light, root cutting, phytotoxic pesticides, high salinity, nutrient deficiencies, and root diseases.

Diagnosis

Eliminate all other possible disorders before considering organic toxins as a cause for plant injury. Conduct a bioassay to determine whether organic toxins are affecting germination: Leach fresh leaves of suspected toxic species with distilled water and use the leachate to water a variety of annual plant seeds planted in soil. Compare the germination of seeds watered with leachate to control seeds that were watered using fresh water.

Remedies

Most organic toxins are short-lived in plant parts and in soil. Rain or irrigation water leaches toxins out of the soil surface.

CHAPTER 6

Case Studies

Case 1: Dead Deodar Cedar *(Cedrus deodara)*

Problem and Symptoms

A mature deodar cedar *(Cedrus deodara)* growing in a cemetery began to decline in mid-August and was completely dead by mid-September (fig. 6.1). The first noticeable symptoms were yellowing of foliage and necrosis of needle tips. Needles progressed from yellow to brown in a few weeks. Symptoms were uniform throughout the tree.

Figure 6.1. This deodar cedar *(Cedrus deodara)* in Modesto, California, began to decline in mid-August and was dead by mid-September. Symptoms included chlorotic and necrotic needles, needle drop, and branch dieback.

Investigation

- Trunk, branches, and foliage were examined for symptoms and signs of bark beetles (exit holes in bark) and other insect pests.
- Soil was removed at the base of the tree to expose the root crown, and the root crown was examined for symptoms of root and crown rot diseases (typical symptoms include trunk cankers, discolored and darkened bark, and brown water-soaked tissues beneath the bark); none were noted. No signs of grade changes (large, basal roots at or below the soil surface) were evident.
- The moisture content and texture of the soil at the tree's drip line were checked at a depth of 3 feet (0.9 m) using an auger (abrupt changes in texture may indicate grade changes and soil drainage problems) and were found to be normal.
- A conversation with the groundskeeper yielded information on cultural practices, including irrigation practices and activities such as gravedigging that may have injured roots.

Possible Causes

Rapid death of an entire tree during hot, dry weather has several possible causes, both biotic and abiotic.

- Irrigation water being applied to the turfgrass may not have been adequate to meet the needs of the tree, and acute water deficit may have occurred, followed by a massive bark beetle attack.
- Root or crown rot diseases (caused by *Phytophthora* spp.) may have rapidly killed the root system, followed by rapid drying and death of the leaves and

branches; frequent irrigation would create wet soil conditions favorable for the development of root rot diseases.

- Raising the soil grade around the tree may have caused a severe aeration deficit.

Figure 6.2. During excavations 2 years prior to decline, over 90 percent of the root system was cut.

Figure 6.3. Being located in a cemetery, the roots had been cut to make room for burial vaults.

- Root cutting or soil compaction may have severely injured the roots.

Diagnosis

The tree was killed by massive soil excavation and root cutting, which removed over 90 percent of the root system (fig. 6.2). Two years earlier, roots were cut and soil was excavated to within 3 feet (0.9 m) of the tree's base and 4 feet (1.2 m) deep for the purpose of installing burial vaults (fig. 6.3). However, the tree did not show symptoms of the injury until nearly 2 years later. In this case, information obtained from the groundskeeper was critical to making an accurate diagnosis.

Treatment and Response

Tree removed. No treatment necessary.

Case 2: Decline of Mixed-Species Planting

Problem and Symptoms

Various species of woody groundcovers, shrubs, and trees in an industrial landscape were either declining or dead (fig. 6.4). The planting was approximately 5 months old. Symptoms included entire and interveinal leaf chlorosis and necrosis, shoot and branch dieback, and death of entire plants. In some species, such as wild lilac (*Ceanothus* spp.), leaf chlorosis and necrosis occurred primarily in older foliage (fig. 6.5). Symptoms were similar in all plants, and all of the plants were affected to some degree (fig. 6.6).

Investigation

- Symptoms over the entire site were similar from species to species; the site was completely free of weeds.
- The moisture content at a depth of 18 inches (0.5 m) near several plants was found to be normal (using a soil sampling tube).
- Soil and water samples were obtained for salinity analysis.
- A conversation with the site manager yielded information on irrigation and

other cultural practices, including extensive chemical weed control programs that had been implemented in the past.

Figure 6.4. A number of species in this landscape were declining or dead.

Figure 6.5. Symptoms on wild lilac (*Ceanothus* sp.) included leaf chlorosis, necrosis, and shoot dieback.

Possible Causes

The uniform decline and death of a large landscape planting consisting of several different species is most likely caused by an abiotic factor.

- Herbicide phytotoxicity: many root-absorbed systemic preemergence herbicides can cause the symptoms observed at the site.
- High soil salinity can also cause leaf chlorosis and necrosis, and in some cases can kill sensitive species.
- Water deficit commonly injures newly planted or poorly established plants.

Diagnosis

Site examinations and visual symptoms indicated that the plants were being injured and killed by a systemic preemergence herbicide. Patterns of leaf chlorosis and necrosis, usually on older foliage first (older leaves had had more time to accumulate the herbicide), are typical of such injury. According to the landscape manager, most of the affected plants were replacements for plants that had died earlier from apparently the same problem. Information obtained from the site manager's files indicated that the herbicides tebuthiuron (Spike), bromacil (Hyvar), and diuron + bromacil (Krovar) had all been used by a commercial applicator to treat the site in previous years, before the current landscape was installed. This information was critical for verifying the visual diagnosis. Laboratory analysis of soil and leaves could have been used to further confirm the diagnosis, but this was considered unnecessary.

Treatment and Response

Recommended treatments included incorporating activated charcoal, compost, or manure into soil to adsorb and remove the herbicides, and using appropriate irrigation practices to improve plant vigor (see "Herbicide Toxicity," p. 188). Due to the expense of treating such a large site, no soil amendments were used. Some plants slowly recovered, but the landscape remains sparse after 10 years.

Figure 6.6. All plants in the landscape were affected and all symptoms were similar.

Figure 6.7. London plane *(Platanus × acerifolia)* trees in this parking lot in San Jose, California, became severely chlorotic within a year after being planted. Trees in adjacent turf planting strips did not exhibit similar symptoms.

Case 3: Chlorotic London Plane (*Platanus acerifolia*)

Problem and Symptoms

London plane trees planted in a commercial parking lot had been declining during the year since they were installed. The trees were planted either in cylindrical concrete planters with a diameter of 30 inches (76 cm) and sidewalls 2 feet (61 cm) deep in asphalt-covered parking areas or in lawn areas between paved areas. The trees in planters exhibited general chlorosis of the canopy, leaf drop, and little or no growth; those in the lawn were asymptomatic (fig. 6.7). All trees were uniformly healthy and vigorous at planting.

Investigation

- An inspection of the site found that trees in the lawn area were irrigated with pop-up sprinklers, while those in planters were irrigated with bubblers (fig. 6.8). The maintenance contractor indicated that the irrigation frequency was high (4 to 5 days per week) in both areas, although the total amount applied was not known. In addition, it was not known whether fertilizer had been applied.
- One tree was removed from the planter (fig. 6.9) and both soil and roots were inspected. The soil was clayey, saturated, blackened, and smelled like rotten eggs (fig. 6.10). Roots appeared water-soaked and discolored. No new roots were found.

Possible Causes

- Aeration deficit: A reduction in air-filled pore space may have occurred due to frequent irrigation, poor drainage, and relatively small soil volume.
- Water deficit is common in parking lot trees. It is often caused by a combination of factors, including infrequent irrigation, high evaporative potential of the site (in this case, reflected light and heat from cars and asphalt), and limited soil volume.

- Salinity could have caused the problem if the trees were irrigated with poor-quality water or overfertilized (either frequently with small amounts of soluble fertilizer or infrequently with large amounts).
- If soil nitrogen was depleted (perhaps by leaching), the soil could have become nitrogen deficient.
- Infection by *Phytophthora* spp. or *Armillaria* spp. may have caused root disease if the soil was maintained in a wet condition for extended periods. The symptoms are consistent with root pathogen infection, but injury is often more severe than that found.

Figure 6.8. Symptomatic trees were planted in cylindrical concrete planters (30-inch [76-cm] diameter) and irrigated with bubblers.

Figure 6.9. Suspecting the cause of injury to be in the root zone, this tree was removed for inspection.

Figure 6.10. Soil in the root zone was clayey, wet, blackened, and smelled like rotten eggs. Roots were watersoaked and blackened. The wire nursery basket had not been removed at planting.

Diagnosis

Based on symptoms and information collected, the most likely cause was determined to be aeration deficit; the symptoms and soil conditions were consistent with this diagnosis. Frequent irrigation, poor drainage (associated with high clay content of the soil), and limited soil volume likely combined to reduce the amount of oxygen available in the root zone. Consequently, root respiration was inhibited, reducing water and mineral uptake. This led to chlorosis, leaf drop, and lack of growth in the canopy. Root disease may have developed as a secondary condition.

Treatment and Response

Improving aeration in the root zone would improve the condition of the trees. This may be achieved by reducing the frequency of irrigation and allowing the soil to dry between irrigations. Due to limited soil volume in the planters, however, this may easily lead to a water deficit.

The relatively small size of the planters makes management of the trees quite difficult. This is a design error that may be corrected by removing the trees and creating larger planting spaces. Soil replacement may be needed as well. If this is not possible, the London plane trees should be replaced with a smaller species that is tolerant of both aeration and water deficits. Unfortunately, using smaller trees will reduce both canopy cover and shading benefit.

Figure 6.11. Severely chlorotic sweetgum *(Liquidambar styraciflua)* trees were performing poorly in an irrigated landscape in Davis, California. Asymptomatic trees were located within 10 yards (9 m) of symptomatic trees.

Figure 6.12. Symptoms included interveinal chlorosis and marginal necrosis of leaves and dieback of shoots. Chlorosis appeared first on youngest leaves.

Case 4: Chlorotic Sweetgum *(Liquidambar styraciflua)*

Problem and Symptoms

In a planting of 20 trees, 3 sweetgum exhibited mild to severe chlorosis (fig. 6.11). Two trees with mild symptoms showed interveinal chlorosis principally on the youngest leaves. In the severely affected tree, chlorosis extended throughout the canopy; the youngest leaves were bleached with marginal necrosis (fig. 6.12). Chlorotic trees were smaller than asymptomatic trees. At planting, all trees were of similar size, age, condition, and stock type (container). Trees were planted in a single row behind a concrete wall and bench and had been in the landscape for 8 years.

Investigation

Samples of soil, water, and leaf tissue were collected for laboratory analysis. Soil samples were taken at two depths, 0 to 3 inches and 3 to 6 inches (0 to 7.5 cm and 7.5 to 15 cm) within the dripline of symptomatic and asymptomatic trees. Tissue samples (leaf blade and petiole) were taken from symptomatic and asymptomatic trees as well. Water samples were collected from an irrigation line serving the planting.

- pH was relatively high for both water (7.7) and soil (7.5 to 7.7), but salt, SAR, chloride, boron, and micronutrient (iron, zinc, manganese) levels in water and soil were neither deficient nor toxic. This was the case for both symptomatic and asymptomatic trees (for soil samples).
- Iron levels in the leaves were the same in chlorotic and nonchlorotic leaves (137 ppm) (fig. 6.13), within a range adequate for most species (see Chapman 1965).
- Manganese (24 ppm for chlorotic leaves, 37 ppm for nonchlorotic) and zinc (27 ppm for chlorotic leaves, 20 ppm for

nonchlorotic) levels were considered not to be deficient.

- Nitrogen content was higher in chlorotic leaves (2.9%) than nonchlorotic leaves (1.86%).
- Landscape maintenance staff reported that neither herbicides nor plant growth regulators had been used on or around the trees.

Possible Causes

- Chlorosis symptoms are typical of micronutrient deficiency, especially iron and/or manganese deficiency. Zinc deficiency may cause similar chlorosis, but it causes the youngest leaves to be abnormally small.
- Soil-applied preemergent herbicides can cause similar symptoms, but the information from the landscape maintenance staff ruled this out.
- Aeration deficit can cause general chlorosis in the canopy, although it is less likely than micronutrient deficiency or chemical injury. Typically, chlorosis caused by aeration deficit is not interveinal and symptoms are not expressed most severely on the youngest leaves.

Figure 6.13. Tissue analysis found that iron (Fe) content in chlorotic and nonchlorotic leaves was the same (137 ppm).

Figure 6.14. An application of iron chelate (200 ppm) sprayed on leaves caused a nonuniform greening response. A uniform greening response may have been achieved by mixing a surfactant with iron chelate. Leaves treated with manganese chelate (200 ppm) and water (controls) did not exhibit a greening response.

Diagnosis

The chlorosis was caused by iron deficiency. Symptoms and laboratory analyses (soil and water) were consistent with this diagnosis. Interveinal chlorosis found in the youngest leaves of the two trees with mild symptoms is typical of iron deficiency. Bleaching and marginal necrosis is typical in more severe cases. As soil pH goes above 7.0, iron becomes less available in soils, causing a deficiency in sensitive species. Field observations have indicated that sweetgum is prone to iron chlorosis in alkaline soils.

This diagnosis was confirmed by field testing using iron chelate (Sequestrene 138). In the spring after leaves had fully expanded, selected chlorotic branches were sprayed to runoff with solutions of iron (200 ppm), manganese (200 ppm), or pure water (control). Leaves on branches treated with iron showed a greening response 3 weeks after treatment (fig. 6.14). Leaves treated with either manganese or water remained chlorotic.

Although this diagnosis was supported by an application of iron chelate, it is unclear why iron content was found to be equivalent in both chlorotic and nonchlorotic leaves. Jacobson and Oertli (1956) reported that an iron deficit can cause a permanent loss of chlorophyll formation in some species, yet iron may still accumulate in leaves. If sweetgum responds to iron deficit in this manner, it would explain the equivalence found in tissue iron content. However, it also begs the question why an iron chelate application would cause a greening response. If chlorophyll loss was permanent, then additional iron should not increase chlorophyll content. Regardless of the mechanism causing iron deficiency symptoms, it is notable that tissue analysis may not be a definitive method for iron deficiency diagnosis in some species. In such cases, iron chelate application would be a more reliable method of diagnosis confirmation.

Treatment and Response

Reduction of soil pH increases iron availability and provides long-term correction of iron chlorosis. This can be achieved by applications of acidifying materials such as sulfur, ferric sulfate, and organic matter. Plant response may be delayed, however, depending on the amount and type of material used and soil conditions, including temperature and moisture content. A more immediate response can be achieved using iron chelate applied to the soil or foliage. However, the effect is usually short-lived (1 year or less), and additional applications are typically needed.

Figure 6.15. Leaf necrosis and branch dieback throughout the canopy of this sweetgum *(Liquidambar styraciflua)* in Los Gatos, California, was noted in early summer. Trees had been treated with ethephon, a plant growth regulator, prior to symptom onset.

Case 5: Canopy dieback in sweetgum *(Liquidambar styraciflua)*

Problem and Symptoms

In early July, several mature sweetgum street trees exhibited signs of canopy dieback and sap flow along the trunk. Leaves on many branches were necrotic, and symptomatic branches were distributed throughout the canopy (figs. 6.15–16). Extensive foaming of a runny exudate was found on large-diameter branches and along the trunk (fig. 6.17). Trees of various ages were affected.

Investigation

- Inspection of neighboring sweetgum street trees showed that some groups of trees were symptomatic while others were asymptomatic (fig. 6.18).
- Consultation with the city arborist revealed that the affected trees had been sprayed (8 weeks prior) with ethephon (Florel) for fruit control. Symptoms appeared 3 weeks after treatment. Wanting to eliminate fruit drop from trees located in front of their property, individual homeowners contracted with a commercial applicator for ethephon treatment. Affected trees had been sprayed similarly in previous years with no negative impacts.

Possible Causes

Although dieback and sap flow symptoms resembled injury from a vascular pathogen, information collected suggested that chemical injury from a plant growth regulator was the most likely cause. Only trees treated with ethephon showed symptoms. All trees not treated were asymptomatic, both in the vicinity of treated trees and elsewhere in the

Figure 6.16. Dieback was largely confined to small-diameter branches (< 2 inches [5 cm] in diameter).

Figure 6.17. Frothing and wetting of the bark was evident on the trunk and large-diameter branches.

city. In addition, symptom onset followed shortly after ethephon treatment.

Diagnosis

Information collected and symptoms point to chemical injury as the cause of dieback. Although ethephon is a plant growth regulator commonly used without injurious effects on a variety of species (including sweetgum) for fruit control, the commercial applicator indicated that other trees treated at the same time in another neighborhood showed similar injury symptoms. However, trees treated in prior weeks showed no effects. Based on this information, the injury probably occurred for one or more of the following reasons: high temperatures at the time of application, an application rate of ethephon exceeding the label rate, or phytotoxic residues in the spray tank.

Treatment and Response

Branches injured by chemical applications cannot be treated. Maintain cultural practices that favor healthy sweetgum develop-

Figure 6.18. Nearby sweetgum trees not treated with ethephon did not show symptoms found on treated trees.

ment; avoid water deficits in particular. Protect affected trees from further injury by other biotic or abiotic agents. Prevent further chemical injury by avoiding applications during high-temperature periods and by ensuring that the label rate is not exceeded and that residual materials do not contaminate the spray tank.

Case 6: Chronic decline of Chinese hackberry *(Celtis sinensis)*

(Courtesy John Lichter, Tree Associates, Winters, CA)

Problem and Symptoms

Chinese hackberry is a relatively large percentage of the tree population in the city of Davis and on the University of California, Davis, campus. Since 1995, a large number of trees have declined and died (fig. 6.19). Two types of symptoms have been found.

Type 1: Wilt-like symptoms

- Mature trees exhibited low vigor (short annual growth increment), poor canopy density and general chlorosis, small leaf size, wilting, dieback, and eventual death (fig. 6.20). The sapwood of roots and the lower trunk was often stained and had a fishlike odor.
- Decay fungi (*Ganoderma* spp.) and inky cap mushrooms (*Coprinus* spp.) were frequently found on or near trees with advanced symptoms.
- Trees usually died within a couple of years of symptom onset, most commonly during the spring when they attempted to leaf out.

Type 2: Iron-chlorosis-like symptoms

- Foliage was chlorotic or bleached (most severely near the end of the shoot), leading to dieback of small to large branches (fig. 6.21).

Figure 6.19. Since 1995, many Chinese hackberry *(Celtis sinensis)* trees in Davis, California, have declined and died. Symptoms include extensive chlorosis and branch dieback.

- Individual limbs exhibited varying degrees of chlorosis or bleaching, leading to a mosaic effect in the canopy.
- To date, only one tree with type 2 symptoms exhibited sapwood staining (noted in trees with wilt-like symptoms), although extensive wood sampling and dissection has not been conducted. Unlike trees with type 1 symptoms, type 2 trees did not die over a short period, although dieback of large branches was found.

Investigation

Pathogen analyses

Laboratory

Tests for viral pathogens (DSRNA analysis), mycoplasmalike organisms (PCR probe), bacteria (electron microscopy), and fungi (isolations) were negative.

Field

One group of healthy Chinese hackberry growing in containers was inoculated with stained wood collected from trees with wilt symptoms; a second group was inoculated with an isolated culture of *Verticillium*; a third control group was inoculated with clean wood. After approximately 1 year, trees inoculated with stained wood did not produce symptoms, but the trees inoculated with *Verticillium* and the control group produced disease symptoms. Tetracycline (Terramycin) was injected into eight symptomatic trees; only one of the eight trees improved following treatment.

Soil and Tissue Analyses

Analysis of soil in the root zone of healthy and symptomatic trees did not reveal significant differences in elements, although one symptomatic tree was deficient in zinc and manganese. Plant tissue analysis revealed symptomatic trees had lower amounts of magnesium, manganese, and zinc compared

Figure 6.20. Poor vigor, wilt, small leaf size, and canopy thinning are typical type 1 symptoms (A). Branch dieback follows canopy thinning (B). Note healthy tree with typical foliage on right.

to leaves from healthy trees. Two foliar treatments with micronutrient sprays did not produce positive results.

Soils sampled in the root zone of five trees with type 2 symptoms and five asymptomatic trees revealed no significant differences in pH or electrical conductivity (EC). However, trees in soil with high moisture content were found to have more severe foliar symptoms.

Soil bulk density measurements were taken at 12 and 24 inches (0.3 and 0.6 m). Trees were in good condition where bulk density was less than 1.55 g/cm^3, while trees in soils with higher densities exhibited a variety of foliar symptoms.

Twenty Chinese hackberry growing in turf on the UC Davis campus representing a range of type 2 symptoms were grouped according to symptom severity. The following treatments were imposed with five trees serving as replicates for each treatment:

- Iron chelate (Sequestar 6% Iron Chelate WDG) was applied over the turf under the canopy of the tree at the rate of 2 pounds per 1,000 square feet.
- Water-jetted holes 3 feet deep on 4-foot spacing (30 to 50 per tree) were dug under the canopy.
- The water-jetted holes as described above filled with a mixture of 3 parts granular soil sulfur to 1 part ammonium sulfate (175 pounds of mixture per tree).
- The control group received no treatment.

Iron chelate and sulfur treatments produced a measurable reduction in foliar symptoms, while the water-jetted trees and the controls showed no change.

Figure 6.21. Whole trees or sections of the canopy become severely chlorotic (Type 2 symptoms). Branches may eventually die back.

Possible Causes

Trees with type 1 symptoms

- Wilt pathogens: Various fungi, such as *Verticillium* spp., can restrict the movement of water in trees by plugging vascular elements, which causes trees to lose vigor, exhibit reduced leaf size, drop leaves, wilt, die back, and eventually die. Type 1 symptoms closely match those that might be found with an aggressive wilt pathogen.
- Root pathogens: Various fungi, such as *Phytophthora* spp., can cause dieback of roots, leading to wilt symptoms. Type 1 symptoms are typical of root rot caused by these fungi.
- Aeration deficit: Most of the symptomatic trees are found in turf, where irrigation is frequent and aeration may be restricted.

Trees with type 2 symptoms

- Micronutrient deficiency: Alkaline (high) pH may reduce iron availability, which can cause chlorosis symptoms.
- Infection by virus or mycoplasmalike organisms can cause these symptoms to appear.

Diagnosis

No definitive diagnosis has yet been made. Having two sets of symptoms and not being able to resolve whether they have a common cause has added a greater level of complexity to this problem. Biotic organisms may be partially responsible for type 1 symptoms, although there is no evidence that virus or mycoplasmalike organisms are involved. Soil compaction, high soil moisture levels, and pH-related micronutrient deficiency appear to be linked to type 2 symptoms. It is likely that multiple factors are involved in both cases. Further study is needed to identify primary and secondary causes. This case provides a reminder that even when a very extensive analysis is conducted, not all problems are successfully diagnosed.

Treatment and Response

Since a definitive diagnosis has not yet been made, treatment recommendations cannot be made with confidence. Based on information collected and field observations, treatment with iron chelate or soil sulfur, reduction of soil compaction, and decreasing irrigation frequency may be beneficial for trees with type 2 symptoms. A root crown examination to assess presence or absence of sapwood staining and wood decay fungi (e.g., *Ganoderma* spp.) is indicated for trees with type 1 symptoms. If found, soil and water management to promote tree vigor and deter pathogen development may produce positive results.

Glossary

For more information on these and other terms see Agrios 1997; Brady 1974; Britton 1995; California Fertilizer Association 1990; Dreistadt 1994; Harris, Clark, and Matheny 1999; Hartmann 1975; Holliday 1998; Kenneth 1960; Matheny and Clark 1991; Morris 1992; Plaster 1992; Raven, Evert, and Eichhorn 1999; Roberts and Boothroyd 1984; Schmidt 1980; Shigo 1979; Sinclair, Lyon, and Johnson 1987; Strand 1999.

abrasion. Wearing, grinding, or rubbing away by friction.

actinomycetes. Formerly, a group of bacteria that form branching filaments; thallobacteria.

acute. Sudden onset or sharp rise of a problem or injury.

adventitious root or bud. Roots or buds produced in situ from newly organized meristems; not preformed.

alkali soil. Soil in which sodium is the primary cation and is present in large enough quantities to adversely affect plant growth. *See* **sodic soil**.

angiosperm. Flowering plants that have seeds enclosed in an ovary.

anoxia. Total lack of oxygen; also, a reduced supply of oxygen to tissues.

asymptomatic. Without symptoms.

bacteria (bacterium, sing.). Single-celled, microscopic, plantlike organisms that do not produce chlorophyll. Most bacteria obtain nitrogen and energy from organic matter; some bacteria cause plant or animal diseases.

ball-and-burlap. Nursery stock in which the plant is dug with soil surrounding the roots and then wrapped with protective material.

bare root. Nursery stock that is grown in the field and sold without soil around the roots.

bark. Nontechnical term applied to all tissues outside the vascular cambium in a woody stem.

bark cracking. Longitudinal split in the stem, involving bark, cambium, and xylem (compare with **growth crack**); may be vertically or horizontally oriented.

biogenic. Resulting from the actions of living organisms; necessary for life processes.

biotic. Alive; pertaining to living organisms.

bleaching. Loss of color or chlorophyll from leaf tissue.

bleeding. Flow of sap from wounds or other injuries; may be accompanied by foul odor.

blister. Elevation of epidermis containing watery liquid.

blotch. Large, irregular-shaped areas on leaves, stems, or shoots.

bridge graft. Repair graft used when the root system of the tree has not been damaged but the bark on the trunk has been injured.

bulk density. Ratio of the mass of water-free soil to its bulk volume, expressed in pounds per cubic foot (lb/ft^3) or grams per cubic centimeter (g/cc^3); sometimes referred to as *apparent density*.

calcareous. Soil or media containing calcium carbonate; a soil alkaline in reaction because of the presence of calcium carbonate.

caliper. Trunk diameter, especially in nursery stock.

callus. Mass of thin-walled undifferentiated cells that develop because of wounding or as a culture on nutrient media.

cambium. Meristematic tissue that gives rise to phloem and xylem.

canker. Necrotic, often sunken lesion on a stem, branch, or twig of a plant.

canopy thinning. Removal of the part of the crown composed of leaves and small twigs by pruning back to lateral branches.

chaparral. Vegetation composed of drought-resistant, stiff evergreen shrubs; also, an ecosystem composed of drought-tolerant, foothill-type vegetation.

chlorophyll. Green pigment found in the chloroplast, important in the absorption of light energy in photosynthesis.

chloroplast. Green cellular structure in the leaves and other parts of the plant that contain chlorophyll and sometimes addi additional pigments.

chlorosis. Yellowing of normally green tissue due to chlorophyll destruction or failure of chlorophyll to form.

chronic. Occurring over a long period of time.

collar. Ring of trunk tissue that surrounds a lateral branch at its point of attachment to the stem.

compartmentalization. Physiological process that creates the chemical and physical boundaries that limit the spread of disease and decay organisms.

compost. Mixture that consists largely of decayed organic matter used for fertilizing and conditioning soil.

crown. Point at or just below the soil surface where the main stem (trunk) joins the roots; also, the topmost limbs on a tree or shrub.

crown thinning. Selective removal of branches to increase light penetration and air movement through the crown.

cuticle. Thin, waxy layer on the outer wall of epidermal cells that consists primarily of wax and cutin.

deficiency. Shortage of substances necessary to plant health.

defoliation. Loss of leaves.

dehydration. Loss of water from plant tissue.

deicing salts. A combination of salts that are spread to keep areas such sidewalks and roadbeds free of ice.

desiccation. Complete drying; occurs in plants when transpiration exceeds moisture absorption, resulting in wilting and damage to plant tissues.

dieback. Progressive death of twigs and small branches, generally from tip to base.

discoloration. In trees, a chemical response to wounding that causes a darkening of the wood. Wood decay generally follows discoloration.

dormancy. In plants, period of inactivity during winter or periods of cold.

EC_e. Electrical conductivity of a soil extract, a physical quantity that measures the readiness with which a medium transmits electricity. Commonly used for expressing the salinity of irrigation water and soil extracts because it can be directly related to salt concentration. It is expressed in deciseimens per meter (dS/m), millisiemens per centimeter (mS/cm), or millimhos per centimeter (mmhos/cm) at 25°C.

ecotype. Locally adapted variant of an organism, differing genetically from other ecotypes.

EC_w. Electrical conductivity of water, a physical quantity that measures the readiness with which a medium transmits electricity. Commonly used for expressing the salinity of irrigation water and soil extracts because it can be directly related to salt concentration. It is expressed in deciseimens per meter (dS/m), millisiemens per centimeter (mS/cm), or millimhos per centimeter (mmhos/cm) at 25°C.

effluent. Waste material such as liquid industrial refuse or sewage that is discharged into the environment.

electrical conductivity (EC). A physical quantity that measures the readiness with which a medium transmits electricity. Commonly used for expressing the salinity of irrigation water and soil extracts because it can be directly related to salt concentration. It is expressed in deciseimens per meter (dS/m), millisiemens per centimeter (mS/cm), or millimhos per centimeter (mmhos/cm) at 25°C.

epicormic. Shoots that arise from latent or adventitious buds; may occur on branches, stems, or from basal suckers.

etiolation. Condition involving increased stem elongation, poor leaf development, and a lack of chlorophyll; found in plants growing in the dark or with greatly reduced amount of light.

evaporative potential. Potential of a site to cause evapotranspiration from plants.

evapotranspiration. Loss of water from a soil by evaporation or from plants by transpiration.

exchangeable sodium percentage (ESP). Degree of saturation of the soil exchange complex with sodium.

exudation. Process by which material passes from within a plant structure to the outer surface or surrounding medium, as in leaf exudate or root exudate.

fasciation. Malformation of plant tissue or stems commonly manifested as enlargement and flattening, as if fused.

fireblight. Destructive disease caused by the bacterium *Erwinia amylovora* that occurs in plants in the pome family; symptoms include a blackening of shoots, as though burned.

flecking. Development of small marks, spots, streaks, or flakes on leaf tissue.
foot-candle. Unit of illumination.
frizzle. In palms, crisp curling, burning, or scorching of fronds from the tip back.

galls. Various abnormal woody growths that occur on tree stems, branches, trunks, and roots. These growths may be caused by bacteria, mycoplasmalike organisms, insects, nematodes, or mistletoe, or they may be abiotic in origin.
girdling. Ring of dead or damaged tissue around the stem or root.
glazing. To give a smooth, glossy surface to a substrate. Soils can become glazed and somewhat impervious to water and roots.
graft (or bud) union. The point where the cut surfaces of two plants join to form a living union.
grafting. Method of plant propagation by attaching a bud or a scion of one plant onto another plant, forming a living union.
growth crack. Longitudinal split in bark due to normal expansion of cambium and xylem (contrast with **bark cracks**); not considered a defect.
gypsum. Common name for hydrated calcium sulfate ($CaSO_4 \cdot 2H_2O$), a mineral used in the fertilizer industry as a source of calcium and sulfur. Gypsum also is used for reclaiming alkali soils in the western United States. Gypsum may reduce the alkalinity of sodic soils by replacing sodium with calcium.

hardening-off. One of several forms of acclimation associated with either decreasing temperatures (increasing tolerance to low temperature) or the transition from nursery environment to the landscape.
herbicide. Agent used to destroy or inhibit plant growth, especially on weeds.
hydrocarbon. Organic compound that consists only of hydrogen and carbon atoms.
hydrozone. Irrigated area consisting of plant materials that have similar water needs.

immobile. Incapable of being moved in plant tissue or soil.
in-arch. Form of repair grafting used when roots have been damaged or injured.
incipient wilt. Onset or beginning stage of plant wilt.
infiltration rate. Rate at which water enters a soil. The rate depends on the soil texture and structure and the amount of water present.
internode. Part of a stem between two nodes.
interveinal. Space between the veins on a plant leaf.

kinked root. Sharply bent root.

lateral root or branch. Secondary root or branch arising from a primary root or branch.
leach. Removal of materials in solution by the passage of water through soil.
lethal. Capable of causing death.
lime. Calcium carbonate ($CaCO_3$).

marginal. Referring to the edge or border of a leaf, as in marginal burning; also, the lower limit of acceptable conditions, as in marginal soil.
mobile. Capable of being moved. Elements such as nitrogen are mobile in soil.
morphology. Form of an organism.
mosaic. Symptom of certain viral diseases of plants characterized by intermingled patches of normal and light-green or yellowish color.
mottle. Irregular pattern of indistinct light and dark areas.
mulch. Material applied onto the soil surface to prevent drying, to prevent rapid changes in soil temperature, as a soil amendment, for decorative purposes, or to prevent weed growth.

necrosis. Death and discoloration of tissue.
node. Part of a stem where one or more leaves are attached.
nonselective herbicide. Herbicide that is phytotoxic to all or most of the plants it comes in contact with.
organelles. Specialized membrane-bound part of a cell.
oxygen diffusion rate (ODR). Rate at which oxygen can be replenished if it is used by respiring plant roots or replaced by water. Expressed as micrograms per square centimeter per minute ($\mu g/cm^2/min$).

pathogen. Entity, usually a microorganism, that can incite disease.

pH. Numerical designation of acidity and alkalinity used for soils and other biological systems; obtained as the common logarithm of the reciprocal of the hydrogen ion concentration of a solution. A pH of 7.0 indicates neutrality; higher values indicate increasing alkalinity, and lower values indicate increasing acidity.

phloem. Conducting tissue that transports organic substances from the mature leaves to other portions of the plant. Translocation usually occurs from areas of high concentration to areas of low concentration.

photosynthesis. Process by which plants convert sunlight into energy.

phytotoxic. Toxic to plants.

pollarding. Pruning technique by which young trees or branches are initially headed and then reheaded on an annual basis without disturbing the callus knob.

postemergence herbicide. Herbicide applied or effective during the period after seedling emergence.

preemergence herbicide. Herbicide applied or effective during the period before seedling emergence.

primary injury. Injury that is the principal or fundamental problem.

rasping. Scraping, as in the rasping mouthparts of an insect.

resin. Solid or semisolid natural organic flammable substance formed by plant secretions that are typically not water-soluble.

root zone. Area of a plant that consists of the roots and the woody and nonwoody tissues that support the crown and absorb water, gases, and nutrients from the soil and atmosphere.

rosette. Short, bunchy habit of plant growth.

saline. Containing soluble salts.

saline-sodic soil. Soil that contains sufficient sodium to interfere with the growth of most crop and landscape plants.

sapwood. Outer and youngest layers of secondary xylem in a trunk or stem that conduct and store water, mineral elements, and carbohydrates.

SAR. Sodium adsorption ratio; ratio that expresses the relative activity of sodium ions in exchange reactions with soil, used with soil extracts and irrigation water. Expressed as milliequivalents per liter (me/L).

saturated paste. Mixture of soil and water commonly used for measurements and for obtaining soil extracts. At saturation the soil paste glistens as it reflects light, flows slightly when the container is tipped, and slides freely and cleanly from a spatula for all soils except those with high clay content.

scaffold. In decurrent trees, the large branches that form the main structure of the crown.

scion. A cutting of a twig or shoot that is grafted onto another twig or shoot.

sclerophyllous. Hard-leaved; leaves resistant to drought by having a great deal of sclerenchymatous (hardened) tissue and reduced intercellular spaces.

scorch. "Burning" of leaf margins caused by infection or unfavorable environmental conditions.

secondary injury. Injury or problem caused by a primary injury, problem, or infection.

secondary root. *See* **lateral root**.

secondary xylem. Growth derived from vascular cambium, including xylem that results in an increase in girth; development of wood.

selective herbicide. Herbicides that are toxic primarily to the target pest (and perhaps a few related species), leaving most other organisms, including natural enemies, unharmed.

senescence. Aging or death.

shoot. The aboveground portions, such as the stem and leaves, of a vascular plant.

silvering. To make plant parts white or silvery, usually as a result of insect feeding.

slipping. Condition in which bark can be easily separated from the wood while the plant is in active growth.

sodic. Media, such as soil, that contains sufficient exchangeable sodium to interfere with the growth of most plants, either with or without appreciable quantities of soluble salts.

softened water. Water treated with sodium to replace calcium and magnesium salts in solution. This process does not lower total salinity.

soil structure. Combination or arrangement of primary soil particles into secondary particles, units, or peds.

soil texture. Relative proportion of the soil separates in a soil; sand, silt, and clay plus the fraction of organic material.

specific ion toxicity. Excess levels of a specific ion (e.g., boron) in plant tissue, causing phytotoxicity.

stippling. Small spots or flecks that create light-colored areas on the leaf. Often caused by insects feeding with piercing-sucking mouthparts.

stock. In grafting, the lower part of the plant containing the root (rootstock).

stomata. Minute openings in the leaves through which gas exchange occurs, including water vapor and carbon dioxide.

stunting. Checking or hindering normal plant growth; to dwarf.

sublethal. Only slightly less than lethal.

sunburn. Injury to bark and cambium caused by a combination of excessive light, heat, and insufficient moisture. *See also* **sunscald**.

sunscald. Injury to bark tissues on the trunk and branches caused by rapid changes in temperature, especially on warm days and cold nights in winter; also, damage to fleshy or herbaceous plant parts as a result of high temperatures other than those caused by direct solar radiation and insufficient moisture (*see* **sunburn**).

symptom. External and internal reaction or alteration of a plant as a result of a disease or injury.

symptomatic. Expressing, or having, symptoms.

temperate. Moderate, as in climate.

tensiometer. Device used to measure the tension with which water is held in the soil.

thrips. Small insect of the order Thysanoptera that has piercing, sucking mouthparts.

toxicity. Capacity of a compound to produce injury.

translocation. Transfer of nutrients (or virus or herbicide) through a plant.

transpiration. Loss of water vapor from the surface of leaves and other aboveground parts of the plant.

tree spikes. Devices that are attached to boots and used by tree workers to aid in tree climbing. Their use inflicts tree wounds.

undifferentiated. Plant tissue that has not developed into roots, stems, buds, cambium, or other identifiable plant organs or structures.

vein clearing. Symptom of virus-infected leaves in which veinal tissue is lighter green that that in normal plants.

vigor. Overall health; the capacity to grow and resist physiological stress.

water soaked. Initial stage of leaf necrosis in spots or irregular-shaped areas. The area turns gray-green as the cells collapse, ultimately drying and forming a lesion.

wet feet. Roots repeatedly or continually exposed to water-saturated soil conditions.

wilt. Loss of rigidity and drooping of plant parts, generally caused by insufficient water in the plant.

winter kill. Plants injured by cold winter temperatures or from becoming deacclimitized on warm days followed by cold nights.

witches' broom. Proliferation of many shoots near the end of a branch in response to the death of the terminal bud.

woody. Plant type that contains secondary xylem.

woundwood. Lignified, partially differentiated tissue that develops from callus associated with wounds.

xeric. Drought-resistant or desert-type plant.

xylem. Plant tissue consisting of tracheids, vessels, parenchyma cells, and fibers; wood.

References

Agrios G. N. 1997. Plant pathology. 4th ed.. New York: Academic Press.

Ahrens, W. H., ed. 1994. Herbicide handbook. 7th ed. Champaign, IL: Weed Science Society of America.

Allen, E. A., D. J. Morrison, and G. W. Wallis. 1996. Common tree diseases of British Columbia. Victoria, BC: Canadian Forest Service, Natural Resources Canada.

Antonelli, A. L., R. S. Byther, R. R. Maleike, S. J. Collman, and A. D. Davison. 1991. How to identify rhododendron and azalea problems. Pullman: Washington State University Cooperative Extension Publication EB1229.

Australian Nature Conservation Agency. 1996. Environment Australia online, biodiversity group. http://www.environment.gov.au/bg/index.htm/

Bailey, L. H. 1935. The standard cyclopedia of horticulture. New York: MacMillan.

Bega, R. B. 1979. Diseases of Pacific Coast conifers. USDA Forest Service Agricultural Handbook 521. Berkeley: Pacific Southwest Forest and Range Experiment Station.

Begeman, J. 1999. Diagnosing palm nutrient deficiencies. Western Arborist 25(4): 42–43.

Bernstein, L. 1958. Salt tolerance of grasses and forage legumes. USDA Information Bulletin 194.

Bernstein, L., L. E. Francois, and R. A. Clark. 1972. Salt tolerance of ornamental shrubs and ground covers. Journal of the American Society of Horticultural Science 97(4): 550–556.

Brady, N. C. 1974. The nature and property of soils. 8th ed. New York: Macmillan.

Brady, N. C., and R. Weil. 1996. The nature and properties of soils. 11th ed. Upper Saddle River, NJ: Prentice-Hall.

Branson, R., and W. Davis. 1965. Ornamental trees and shrubs showing adaptation characteristics suitable for San Mateo County. Unpublished manuscript. San Mateo: Agricultural Extension Service, University of California.

Brenzel, K. N., ed. 1998. Sunset western garden problem solver. Menlo Park, CA: Sunset.

———. 2001. Sunset western garden book. 7th ed. Menlo Park, CA: Sunset.

Brickell, C., and J. Elsley. 1992. Encyclopedia of garden plants. New York: Macmillan.

Britton, J. C. 1995. Tree pruning guidelines. Savoy, IL: International Society of Arboriculture.

Bunt, A. C. 1976. Modern potting composts. University Park: Pennsylvania State University Press.

Butin, H. 1995. Tree diseases and disorders. Oxford: Oxford University Press.

Byther, R. S., C. R. Foss, A. L. Antonelli, R. R. Maleike, and V. M. Bobbit. 1996. Landscape plant problems: A pictorial diagnostic manual. Tacoma, WA: Pollard Group.

———. 2000. Landscape plant problems: A pictorial diagnostic manual. 2nd ed. Pullman: Washington State University.

California Air Resources Board. 1984. The effect of smog on California plants. Sacramento: Air Resources Board.

California Fertilizer Association. 1990. Western fertilizer handbook, horticulture edition. Sacramento: California Fertilizer Association.

California State Polytechnic University, Pomona. N.d. Soil materials syllabus, SS 439. Unpublished manuscript.

Cardon, G. E., and J. J. Mortvedt. 1999. Salt-affected soils. Colorado State University Cooperative Extension. http://www.colostate.edu/Depts/CoopExt/PUBS/CROPS/00503.html/

Cathey, H. M. 1998. Heat-zone gardening. Alexandria, VA: Time-Life Books.

Chapman, H. D. 1965. Diagnostic criteria for plants and soils. Berkeley: University of California Division of Agricultural Sciences.

Chase, A. R., and T. K. Broschat, eds. 1991. Diseases and disorders of ornamental palms. St. Paul, MN: APS Press.

Cornell Cooperative Extension. 1999. Dying of thirst (salt damage in landscapes). The Horticulture Lookout 1(10): 9–11.

Costello, L. R., and K. S. Jones. 2000. A guide to estimating the irrigation water needs of landscape plantings in California: WUCOLS III. Sacramento: California Department of Water Resources.

Craul, P. J. 1992. Urban soil in landscape design. New York: Wiley.

Cripe, R. E. 1978. Lightning protection for trees. American Nurseryman 148(9): 58–60.

D'Ambrosio, D. P. 1990. Crown density and its correlation to girdling root syndrome. Journal of Arboriculture 16:153–57.

Derr, J. F., and B. L. Appleton. 1988. Herbicide injury to trees and shrubs: A pictorial guide to symptom diagnosis. Virginia Beach: Blue Crab Press.

Dirr, M. A. 1978. Tolerance of seven woody ornamentals to soil-applied sodium chloride. Journal of Arboriculture 4(7): 162–165.

———. 1997. Dirr's hardy trees and shrubs: An illustrated encyclopedia. Portland, OR: Timber Press.

Donaldson, D. R., J. K. Hasey, and W. B. Davis. 1983. Eucalyptus out-perform other species in salty, flooded soils. California Agriculture 37(9–10): 20–21.

Donselman, H. Palm nutrition. Western Arborist 23(2): 7–9.

Downer, A. J. 1993. The truth about eucalyptus. Landscape Notes Newsletter 8(1):7–10.

Dreistadt, S. H. 1994. Pests of landscape trees and shrubs: An integrated pest management guide. Oakland: University of California Division of Agriculture and Natural Resources Publication 3359.

Driver, C. 1990. Planting in hot arid climates. In B. Clouston, ed., Landscape design with plants. Boca Raton, FL: CRC Press.

Eaton, F. M. 1935. Boron in soils and irrigation waters and its effect on plants with particular reference to the San Joaquin Valley of California. USDA Technical Bulletin 448. Washington, DC: USDA.

———. 1944. Deficiency, toxicity and accumulation of boron in plants. Journal of Agricultural Research 69:237–277.

EBMUD (East Bay Municipal Utility District). 1990. Water-conserving plants & landscapes for the Bay Area. Oakland, CA: EBMUD.

Elmore, C. L. N.d. Symptoms of herbicide injury. Davis: University of California, Davis, Weed Science Program.

Farnham, D. S., R. S. Ayers, and R. F. Hasek. 1985. Water quality: Its effect on ornamental plants. Oakland: University of California Division of Agriculture and Natural Resources Leaflet 2995.

———. 1998. A property owner's guide to reducing the wildfire threat. Oakland: University of California Division of Agriculture and Natural Resources Leaflet 21539.

Flagler, R. B. 1998. Recognition of air pollution injury to vegetation: A pictorial atlas. 2nd ed. Pittsburgh, PA: Air and Waste Management Association.

Francois, L. E. 1980. Salt injury to ornamental shrubs and ground covers. USDA Home and Garden Bulletin 231.

———. 1982. Salt tolerance of eight ornamental tree species. Journal of the American Society of Horticultural Science 107(1): 66–68.

Francois, L. E., and R. A. Clark. 1978. Salt tolerance of ornamental shrubs, trees and iceplant. Journal of the American Society of Horticultural Science 103(2): 280–283.

———. 1979. Boron tolerance of twenty-five ornamental shrub species. Journal of the American Society of Horticultural Science 104(3): 319–322.

Gilman, E. F. 1997. Trees for urban and suburban landscapes. Albany, NY: Delmar.

Gilmer, M. 1997. The myth and truth of fire-resistant plants. California Landscaping (April): 10–13.

Glattstein, J. 1989. Ornamentals for sandy and saline soils. Grounds Maintenance (April): 52–60.

Green, J. L., O. Maloy, and J. Capizzi. 1990. A systematic approach to diagnosing plant damage. Available at Oregon State University's ornamentals Web site: http://osu.orst.edu/dept.hort/dpd/nindex.htm

Hagen, B. W. 2000. Redwoods damaged by high temperatures. Western Arborist 26(2): 17.

Harivandi, A. 1988. Irrigation water quality and turfgrass management. California Turfgrass Culture 38(3/4): 1–4.

Harris, R. W., J. R. Clark, and N. P. Matheny. 1999. Arboriculture: Integrated management of landscape trees, shrubs and vines. 3rd ed. Upper Saddle River, NJ: Prentice-Hall.

———. 2003. Arboriculture: Integrated management of landscape trees, shrubs and vines. 4th ed. Upper Saddle River, NJ: Prentice-Hall.

Harris, R. W., J. L. Paul, and A. T. Leiser. 1977. Fertilizing woody plants. Oakland: University of California Agricultural Sciences Leaflet 2958.

Hartman, H. T., and D. E. Kester. 1975. Plant propagation: Principles and practices. 3rd ed. Englewood Cliffs, NJ: Prentice-Hall.

Hartman, J. R., T. P. Pirone, and M. A. Sall. 2000. Pirone's tree maintenance. 7th ed. New York: Oxford University Press.

Hartmann, K. 1975. Plant propagation. 3rd ed. Upper Saddle River, NJ: Prentice-Hall.

Hesketh, K. A., and G. W. Hickman. 1984. Organic soil amendments for home gardening. In H. Johnson and D. Pittenger, eds., Vegetable briefs no. 236. Oakland: University of California Division of Agriculture and Natural Resources.

Holliday, P. 1998. Dictionary of plant pathology. 2nd ed. Cambridge, UK: Cambridge University Press.

Hoyt, R. S. 1978. Ornamental plants for subtropical regions. Anaheim, CA: Livingston Press.

Hume, E. 1990. Landscaping for defensible space. Proceedings of Statewide Wildfire Conference, Section 2. Ames: Iowa State University Press.

Jacobson, L., and J. J. Oertli. 1956. The relation between iron and chlorophyll contents in chlorotic sunflower leaves. Plant Physiology 31:199–204.

Jarrell, W. M., and T. Furuta. N.d. Evaluation of eucalyptus materials as ingredients of composts and potting mixes. Final Report, Elvenia J. Slosson Fund for Ornamental Horticulture. Oakland: University of California Division of Agriculture and Natural Resources.

Jenkins, W. R., and D. P. Taylor. 1967. Plant nematology. New York: Reinhold.

Johnson, G. R., and E. Sucoff. 1995. Minimizing de-icing salt injury to trees. Minneapolis: Univ. of Minnesota Ext. Service FO-1413-GO. http://www.extension.umn.edu/distribution/naturalresources/DD1413.html/

Kenneth, J. A. 1960. A dictionary of scientific terms. 7th ed. Princeton: Van Nostrand.

Kluepfel, M., J. M. Scott, J. H. Blake, and C. S. Gorsuch. 1999. Galls and outgrowths. Clemson, SC: Clemson University Cooperative Extension Fact Sheet 2352.

Kocher, S., R. Harris, and G. Nakamura. 2001. Recovering from wildfire: A guide for California's forest landowners. Oakland: University of California Division of Agriculture and Natural Resources Publication 21603.

Kopinga, J., and J. van den Burg. 1995. Using soil and foliar analysis to diagnose the nutritional status of urban trees. Journal of Arboriculture 21(1): 17–24.

Kozlowski, T. T., ed. 1984. Flooding and plant growth. Orlando, FL: Academic Press.

Kramer, P. J., and J. S. Boyer. 1995. Water relations of plants and soils. Orlando, FL: Academic Press.

Kuhns, J., and L. Rupp. 2000. Selecting and planting landscape trees. Utah State University Extension NR-460. http://extension.usu.edu/publica/natrpubs/nr460.pdf

Kvaalen, R. N.d. Roadside ornamental plants and de-icing salts. Unpublished manuscript. West Lafayette, IN: Purdue University Cooperative Extension.

Labanauskas, C. K. 1966. Aluminum. In H. D. Chapman, ed., Diagnostic criteria for plants and soils. Berkeley: University of California Division of Agricultural Sciences. 264–285.

Laemmlen, F. 1980. Diagnosis of problems in trees and shrubs. Appleton, WI: Madison Publishing.

Lange, A. H., C. L. Elmore, and A. B. Saghir. 1969. Diagnosis of phytotoxicity from herbicides in soils. Oakland: University of California Cooperative Extension Leaflet TA-69.

Leone, I. A., E. F. Gilman, M. F. Telson, and F. B. Flower. 1980. Selection of trees and planting techniques for former refuse landfills. Metropolitan Tree Improvement Alliance (METRIA) Proceedings 3:107–117.

Lichter, J. M., and L. R. Costello. 1994. An evaluation of volume excavation and core sampling techniques for measuring soil bulk density. Journal of Arboriculture 20(3): 160–164.

Litzow, M., and H. Pellett, 1983. Materials for potential use in sunscald prevention. Journal of Arboriculture 9(2): 35–38.

Lubin, D. M. and J. Shelly. 1997. Defensible space landscaping in the urban/wildland interface. Richmond: University of California Forest Products Laboratory Internal Report No. 36.01.137. Available online at http://www.ucfpl.ucop.edu/I-Zone/XIV/vegetati.htm#Favorable

Maas, E. V. 1984. Salt tolerance of plants. In B. R. Christie, ed., The handbook of plant science in agriculture. Boca Raton, FL: CRC Press.

MacKenzie, J. J., and M. T. El-Ashry. 1989. Air pollution's toll on forest and crops. New Haven: Yale University Press.
Maire, R., 1976. Landscape for fire protection. Oakland: University of California Division of Agriculture and Natural Resources Leaflet 2401.
Martin-Richardson, B. N.d. A homeowner's guide to fire resistant plants for the San Luis Obispo area. San Luis Obispo: California Department of Forestry and San Luis Obispo County Fire Department.
Matheny, N. P., and J. R. Clark. 1991. A photographic guide to the evaluation of hazard trees in urban areas. Urbana, IL: International Society of Arboriculture.
———. 1994. A photographic guide to the evaluation of hazard trees in urban areas. 2nd ed. Champaign, IL: International Society of Arboriculture.
———. 1998. Managing landscapes using recycled water. In D. Neely and G. W. Watson, eds., The landscape below ground II. Champaign, IL: International Society of Arboriculture.
McCain, A. H., and D. R. Donaldson. 1984. Western gall rust. Oakland: University of California Division of Agriculture and Natural Resources Leaflet 2431.
McClintock, E., and A. T. Leiser. 1979. An annotated checklist of woody ornamental plants of California, Oregon, and Washington. Oakland: University of California Division of Agriculture and Natural Resources Publication 4091.
McMinn, H. E. 1959. An illustrated manual of California shrubs. Berkeley: University of California Press.
Michigan State University Extension. 1996. Chemical injury on evergreens. http://msue.msu.edu/msue/imp/mod03/01701168.html/
Millar, C. E., L. M. Turk, and H. D. Foth. 1972. Fundamentals of soil science. 5th ed. New York: Wiley.
Miller, F. 2000. Want to be a better diagnostician? Tips for proper plant diagnosis. Arborist News 9(4): 33–38.
Mills, H. A., and J. B. Jones Jr. 1996. Plant analysis handbook II. Athens, GA: MicroMacro Publishing.
Moje, W. 1966. Organic soil toxins. In H. D. Chapman, ed., Diagnostic criteria for plants and soils. Berkeley: University of California Division of Agricultural Sciences.
Moritz, R., and P. Svihra. 1996. Pyrophytic vs. fire resistant plants. Hortscripts No. 18. Novato: University of California Cooperative Extension, Sonoma and Marin Counties.
Morris, C., ed. 1992. Academic Press dictionary of science and technology. San Diego: Academic Press.
Morris, L., and D. Devitt. 1990. Salinity and landscape plants. Grounds Maintenance (April): 6, 8.
———. 2001. Soil temperature and root growth. Western Arborist 27(3): 38–39.
Noble , R. D., J. L. Martin, and K. F. Jensen. 1988. Air pollution effects on vegetation, including forest ecosystems. Proceedings of the Second US–USSR Symposium, Corvallis, OR, 1988. Gov. Doc. A13.42/2:2:A:7. Bromall, PA: Northeastern Forest Experiment Station.
Partyka, R. E. 1982. The ways we kill a plant. Journal of Arboriculture 8(3): 57–66.
Pennsylvania State University. 1987. Diagnosing injury to eastern forest trees. State College: Pennsylvania State University College of Agriculture Sciences.
Perry, B. 1989. Trees and shrubs for dry California landscapes. Claremont, CA: Land Design Publishing.
Perry, E., and G. W. Hickman. 2001. A survey to determine the leaf nitrogen. Journal of Arboriculture 27:3.
Pettygrove, G., and T. Asano. 1985. Irrigation with reclaimed municipal wastewater—A guidance manual. Chelsea, MI: Lewis Publishers.
Pirone, P. P. 1978. Diseases and pests of ornamental plants. 5th ed. New York: Wiley.
Pirone, P. P., et al. 1988. Tree maintenance. 6th ed. New York: Oxford University Press.
———. 1999. Diseases and pests of ornamental plants. 7th ed. New York: Wiley.
Pittenger, D., ed. 2002. California master gardener handbook. Oakland: University of California Division of Agriculture and Natural Resources Publication 3382.
Plaster, E. J. 1992. Soil science and management. 2nd ed. New York: Delmar.
Plumb, T. R., and A. P. Gomez. 1983. Five Southern California oaks: Identification and post-fire management. General Technical Report PSW-71. Berkeley: USDA Forest Service Pacific Southwest Range and Experiment Station.

Pratt, P. F. 1966a. Aluminum. In H. D. Chapman, ed., Diagnostic criteria for plants and soils. Berkeley: University of California Division of Agricultural Sciences. 3–12.

———. 1966b. Carbonate and bicarbonate. In H. D. Chapman, ed., Diagnostic criteria for plants and soils. Berkeley: University of California Division of Agricultural Sciences. 93–97.

Questa Engineering Corporation. 1987. Irrigation water quality study, City of Concord. Unpublished manuscript. Point Richmond, CA: Questa.

Raabe, R. D., A. H. McCain, and A. O. Paulus. 1977. Diseases of camellias in California. Oakland: University of California Division of Agriculture and Natural Resources Leaflet 2151.

Raven, P. H., R. F. Evert, and S. E. Eichhorn. 1999. Biology of plants. 6th ed. New York: Worth.

Reisenauer, H. M., ed. 1978. Soil and plant-tissue testing in California. Oakland: University of California Division of Agriculture and Natural Resources Bulletin 1879.

Reuther, W., and C. K. Labanauskas. 1966. Copper. In H. D. Chapman, ed., Diagnostic criteria for plants and soils. Berkeley: University of California Division of Agricultural Sciences. 157–179.

Roberts, D. A., and C. W. Boothroyd. 1984. Fundamentals of plant pathology. 2nd ed. New York: Freeman.

Roewekamp, F. W. 1941. Boron in Los Angeles water supply in relation to ornamental plants. Unpublished paper prepared for Horticulture 281. University of California, Los Angeles, library.

Roppolo, D. J., Jr., and R. W. Miller. 2001. Factors predisposing urban trees to sunscald. Journal of Arboriculture 27:5.

San Diego, City of. 1966. Water reclamation study for Balboa Park and Mission Bay Park. Unpublished manuscript. San Diego: Boyle Engineering.

Schmidt, M. G. 1980. Growing California native plants. Berkeley: University of California Press.

Shigo, A. L. 1979. Tree decay: An expanded concept. USDA Forest Service Agricultural Information Bulletin 419.

Shurtleff, M. C. 1966. How to control plant diseases in home and garden. Ames: Iowa State University Press.

Sinclair, W. A., H. H. Lyon, and W. T. Johnson. 1987. Diseases of trees and shrubs. Ithaca: Cornell University Press.

Singer, M. J., and D. N. Munns. 1987. Soils: An introduction. New York: Macmillan.

Skimina, C. A. 1980. Salt tolerance of ornamentals. International Plant Propagation Society 30:113–118.

Skroch, W. A., and T. J. Sheets, eds. 1985. Herbicide injury symptoms and diagnosis. Raleigh: North Carolina Agricultural Extension Service.

Standiford, R. B. 1991. Assessing fire-damaged coastal live oaks. Berkeley: University of California Berkeley Department of Forestry and Resources Management.

Strand, L. L. 1999. Integrated pest management for stone fruits. Oakland: University of California Division of Agriculture and Natural Resources Publication 3389.

Svihra, P., and B. D. Coate. 1996. Landscape plants that may survive adverse soil and water conditions. HortScript No. 19. Novato: University of California Cooperative Extension Tickes, B., D. Cudney, and C. Elmore. 1996. Herbicide injury symptoms. Tucson: University of Arizona Cooperative Extension Publication 195021.

Treshow, M. 1970. Environment and plant response. New York: McGraw-Hill.

USDA plant hardiness zone map. 1990. USDA Miscellaneous Publication 1475. Washington, D.C.: Government Printing Office. Also available online at www.usna.usda.gov/Hardzone/

Van Arsdel, E. P. 1980. Managing trees to reduce damage from low-level saline irrigation. Weeds Trees & Turf (June): 26–28, 61.

Wysong, D. S., M. O. Harrell, and D. H. Steinegger. 2000. Iron chorosis of trees and shrubs. NebGuide G94-1218-A. Lincoln: Cooperative Extension, Institute of Agriculture and Natural Resources, University of Nebraska-Lincoln. http://www.ianr.unl.edu/pubs/plantdisease/g1218.htm

Wu, L., X. Guo, A. Harivandi, R. Waters, and J. Brown. 1999. Study of California native grass and landscape plant species for recycled water irrigation in California landscapes and gardens. Oakland: University of California Division of Agriculture and Natural Resources.

Wu, L., E. Zagory, R. Waters, and J. Brown. 2001. Tolerance of landscape plants to recycled water irrigation. Growing Points 5(4).

Index

Note: Page numbers in **boldface type** indicate major discussions. Page numbers in *italic type* indicate photographs or illustrations. Page numbers followed by *t* indicate tables.

F

G